PLATE TECTONICS

PLATE TECTONICS

UNRAVELING THE MYSTERIES OF THE EARTH

The Changing Earth Series

JON ERICKSON

Facts On File

New York • Oxford

PLATE TECTONICS:
UNRAVELING THE MYSTERIES OF THE EARTH

Copyright © 1992 by Jon Erickson

Facts On File, Inc.
460 Park Avenue South
New York NY 10016
USA

Facts On File Limited
c/o Roundhouse Publishing Ltd.
P.O. Box 140
Oxford OX2 75F
United Kingdom

Library of Congress Cataloging-in-Publication Data
Erickson, Jon, 1948–
   Plate tectonics : unraveling the mysteries of the earth / Jon Erickson.
     p. cm. —(The changing earth)
   Includes bibliographical references and index.
   ISBN 0-8160-2588-6 (acid-free paper)
   1. Plate tectonics. 2. Geobiology. 3. Astrogeology. I. Title.
   II. Series: Erickson, Jon, 1948– Changing earth.
   QE511.4.E75 1992
   551.1'36—dc20   91-34391

A British CIP catalogue record for this book is available from the British Library.

Text design by Ron Monteleone
Jacket design by Catherine Hyman
Composition by Facts On File, Inc.
Manufactured by R.R. Donnelley & Sons, Inc.
Printed in the United States of America

10 9 8 7 6 5 4 3 2 1

This book is printed on acid-free paper.

CONTENTS

TABLES IN PLATE TECTONICS

ACKNOWLEDGMENTS

The author wishes to thank the following organizations for providing photographs for this book: the National Aeronautics and Space Administration (NASA), the National Oceanic and Atmospheric Administration (NOAA), the U.S. Air Force, the U.S. Department of Agriculture–Forest Service, the U.S. Geological Survey (USGS), and the U.S. Navy.

INTRODUCTION

All aspects of the Earth's history and structure, including its majestic mountains, giant rift valleys, and deep ocean basins, were fashioned by mobile tectonic plates (from the Greek *tekton*, meaning "to build"). The Earth's outer shell comprises a dozen or so rigid plates composed of the upper mantle, or lithosphere, and the overlying crust, that are always in motion. This movement accounts for all geologic activity taking place on the Earth's surface. The plates ride on the semimolten rocks of the Earth's mantle and carry the continents along with them like ships frozen in arctic pack ice.

When two plates collide, they create mountain ranges on the continents or volcanic islands on the ocean floor. When an oceanic plate slides beneath a continental plate, it produces sinuous mountain chains, such as the Andes of South America, and volcanic mountain ranges, such as the Cascades of the western United States. The breakup of a plate creates new continents and oceans. The process of rifting and patching of the continents has been ongoing for the past 2.7 billion years and possibly even longer.

Along the middle of the Atlantic Ocean runs an impressive submarine mountain range called the Mid-Atlantic Ridge, which surpasses in scale the Alps and Himalayas combined. It is part of a global midocean ridge system that stretches over 40,000 miles along the ocean floor like the seam on a baseball. A deep trough is carved down the middle of the ridge as though it were a giant crack in the Earth's crust. The rift is four times deeper than the Grand Canyon, making it the longest and deepest canyon on the planet.

The Mid-Atlantic Ridge is the center of intense seismic and volcanic activity and the focus of high heat flow from the Earth's interior. Molten magma from deep inside the mantle rises and erupts on the ocean floor, adding new oceanic crust to both sides of the ridge crest. In the meantime, the upwelling magma pushes apart the two plates upon which the continents surrounding the Atlantic ride.

Currently the Atlantic Basin is widening and pushing North America and Europe apart along the Mid-Atlantic Ridge at a rate of about an inch per year. As the plates spread apart in the Atlantic, the ocean floor of the Pacific is destroyed to make more room. The Pacific Basin is ringed with zones where these plates sink into the mantle when they become too thick and heavy to remain on the surface. As the plate is descending, it drags the rest of the ocean floor along with it like a locomotive pulling a freight train.

On its journey deep into the mantle, the crust melts, and the molten rocks rise toward the surface. When they reach the base of the continental crust, they become the source of a new magma for volcanoes. The magma also erupts on the ocean floor, forming long chains of volcanic islands.

In this manner, plate tectonics is continuously changing and rearranging the face of the Earth.

1

CONTINENTAL DRIFT

When early cartographers drew up the first world maps, they noticed something very peculiar. The continents of Africa and South America seemed to fit together as though they were pieces of a giant jigsaw puzzle. Many other clues, such as matching mountain ranges and identical life forms on continents widely separated by ocean, led to speculation that at one time all the lands were united into a single large continent that subsequently broke apart. This was a preposterous notion to most scholars, who then devised all sorts of theories to explain these strange phenomena.

BIRTH OF GEOLOGY

During the late Renaissance, a rebirth of scientific inquiry into natural phenomena took hold of Europe after nearly 1000 years of silence during the Middle Ages. Around the beginning of the 18th century, the French chemist Nicolas Lemery delved into why volcanoes erupted. He observed how a mixture of iron filings, sulfur, and water spontaneously combusted and gave off steam and hot projected matter. In Lemery's view, sulfur fermented in the depths of the Earth to produce earthquakes and confla-

MOUNT VESUVIUS,
Italy, 1944

Figure 1 Eruption of Mount Vesuvius in 1944. Courtesy U.S. Geological Survey (USGS)

grations. He concluded that volcanoes were the product of fermentation and combustion of certain matter when they came in contact with air and water.

The French naturalist Georges de Buffon championed this idea and concluded that the center of volcanic activity was not deep in the bowels of the Earth but near the surface, where it was exposed to wind and rain. This theory became known as neptunism, named for Neptune, the Roman god of the sea. The German geologist Abraham Werner became one of its prime exponents and maintained that once the mineral pyrite was exposed to water it heated up and ignited coal. The burning coal melted nearby rocks, which then erupted on the Earth's surface.

The Scottish geologist James Hutton, known today as the father of modern geology, argued against neptunism and advanced an opposing theory called plutonism, named for Pluto, the Greek god of the underworld. His theory was based on the premise that the depths of the Earth are in constant turmoil and that molten matter rises to the surface through cracks or fissures, giving rise to an erupting volcano. However, it was not until the middle 19th century when the first serious study of the Mount Vesuvius Volcano near Naples, Italy (Figure 1), was conducted that conclusive evidence was found supporting the theory of plutonism.

Other theories underpinning the foundation of modern geology also developed over time. The 17th-century Italian physician Nicholas Steno recognized that in a sequence of layered rocks, undeformed by folding or faulting, each layer was formed after the one below it and before the one above it. This led Steno to propose the law of superposition, which might seem obvious to us today, but during his time it was hailed as an important scientific discovery. Steno also put forward the principle of original horizontality, which states that sedimentary rocks were initially laid down horizontally in the ocean, and subsequent folding and faulting uplifted them out of the sea and inclined them at steep angles.

If angled rocks are overlain by horizontal strata, there existed a gap in the chronological record known as an angular unconformity (Figure 2). Also, if a body of igneous rocks, rocks derived from molten magma, cuts across the boundaries of other rock units, it is younger than those it intercepts. This is the principle of cross cutting relationship, which states that granitic intrusions, caused by the injection of magma into pre-existing rocks, are younger than the rocks they invade. A sequence of rocks placed in their proper order is called a stratigraphic cross section.

The development of a geologic time scale applicable around the world required that rock units from one locality be matched with rock units of similar age in another locality. This is called rock correlation. The correlation of rocks from one place to another over a wide area therefore helps to establish the geologic history of the entire region. Geologists can trace a bed or a series of beds from one outcrop to another by recognizing certain distinguishable features in the rocks.

Matters are complicated, however, if there is faulting in the area. One block of a rock sequence might be down-dropped in relation to the other (Figure 3) or thrusted on top of another. Rocks that occur in repetitive sequences of sandstone, shale, and limestone complicate correlation even further. Although these methods might be sufficient for tracing rock forma-tions over relatively short distances, they are inadequate for matching rocks over great distances, such as between continents.

In order to correlate between widely sepa-rated areas, geologists had to rely on the fossil content of the rocks. A fossil is the remains or traces of an organism preserved from the geo-

Figure 2 Angular unconformity in small mesa near San Lorenzo, Socorro County, New Mexico. Photo 204 by R. H. Chapman, courtesy USGS

logic past. A plant or an animal must be buried rapidly in the absence of oxygen or bacteria to prevent decay. Given enough time, the remains of an organism are modified, often becoming petrified and literally turned into stone. The branch of geology devoted to the study of ancient life based on fossils is called paleontology.

Although the existence of fossils has been known since the ancient Greeks, it was not until the 18th century that their significance as a geologic tool was discovered. The British engineer and geologist William Smith found that in the canals he built across Great Britain rock strata contained fossils that were different from those in beds either below or above. He also noticed that sedimentary rocks in widely separated areas could be identified by their distinctive fossil content. These observations led Smith to one of the most important and basic principles of historical geology. Fossilized organisms succeed one another in a definite and determinable order. Therefore, various geologic time periods could be recognized by their distinctive fossils.

Recognition of the slow processes by which geologic forces operate led James Hutton to propose the theory of uniformitarianism in 1785. Simply stated, it means that the present is the key to the past. In other words, the forces that shaped the

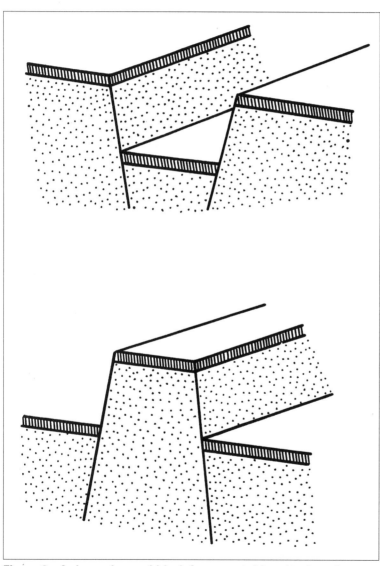

Figure 3   A down-dropped block forms a grabben (top), and an upthrust block forms a horst (bottom).

Earth are uniform and operate in the same manner and at the same rate today as they did in the past. Hutton envisioned the prime mover behind these slow changes to be the Earth's own internal heat. Geologists had long ago recognized that rocks were molten in the Earth's interior, and this was manifested by volcanic eruptions (Figure 4).

Moreover, temperatures in deep mines increased with greater depth, indicating that rocks grew progressively hotter toward the center of the Earth. Hutton called this phenomenon the Earth's great heat engine. He believed that the heat was left over from an earlier time when the planet was in a molten state. The British geologist Sir Charles Lyell, born in 1797, the same year that Hutton died, took up where Hutton left off and gained worldwide acceptance for the theory of uniformitarianism.

Many geologists, however, felt that this theory

Figure 4   The November 1968 eruption of Cerro Negro in west-central Nicaragua, which resembles a chain of volcanoes similar to the Cascade Range of the Pacific Northwest. Photo from Earthquake Information Bulletin 464, courtesy USGS

was not fully adequate to explain all the geologic forces at work. There appeared to be events in the past that were not slowly evolving but occurred rather suddenly. This opposing view was called catastrophism. Its adherents pointed to gaps in the geologic record and the extinctions of large numbers of species. It was thought that the Earth underwent periods of catastrophic death of all life, after which life began anew. This also explained the abundance of fossils at certain stages in the geologic record.

MATCHING COASTLINES

Early geographers were often puzzled over the way the bulge on the east coast of South America fitted almost precisely into the cleft on the west coast of Africa. In 1620, the British philosopher Sir Francis Bacon noticed similarities between the New World and the Old World. Isthmuses and capes looked much the same, and both continents were broad and extended toward the North but narrow and pointed toward the South. He also noticed that the Atlantic coastlines of South America and Africa seemed to match. The French naturalist Georges de Buffon suggested that Europe and North America had once been joined because of similarities between their plants and animals.

The French moralist François Placet suggested that the New and Old Worlds were once joined together and then separated by the Great Flood during Noah's time. Geologists in the 17th and 18th centuries argued that the Flood was so devastating that it broke up old continents into entirely new ones. Expanding on this idea, the 19th-century German naturalist-explorer Alexander von Humboldt suggested that a massive tidal wave surged across the globe and carved out the Atlantic Ocean like a giant river valley, leaving the continents divided with opposing shorelines. Humboldt studied the landforms and plant and animal life in South America and noticed that mountain ranges on the eastern coast resembled those on the western coast of Africa. There were striking resemblances in the geologic strata of the two continents as well.

In the mid-19th century, the American writer Antonio Snider suggested that as the Earth cooled and crystallized from a molten state most of the continental material gathered on one side, making the planet lopsided and unbalanced. This created such internal stresses that a large continent cracked wide open, and hot lavas bled through the fissures. Meanwhile, the rains of the Great Flood came, and the raging waters pushed apart the segments of the broken continent to their present positions. Snider cited as evidence for a single large landmass coal beds in Africa and South America that were the same age and contained similar fossils.

A popular theory for the creation of the moon contended that it was torn out of the Earth by the tidal pull of a passing star. In 1882, the British scientist Osmond Fisher suggested that when the moon was plucked out of the Earth it left a great scar that formed the Pacific Basin, which quickly filled with molton magma from the Earth's interior. As the upper fluid layers flowed into this cavity, the cooling solid crust floating on top was broken up, and a portion of it was pulled toward the cavity like a raft riding on a river of molten rock.

In 1885, the Austrian geologist Edward Suess demonstrated how the continents of the Southern Hemisphere were united into a composite landmass he called Gondwanaland (Figure 5), after a province in India.

Suess named the northern landmass Laurasia, for the Laurentian province of Canada and the continent of Eurasia. During Suess's day, geologists thought that as the Earth was cooling down it shriveled up like a baked apple, accounting for the formation of mountain belts.

Near the turn of the 20th century, biologists and geologists alike were bewildered by the similarity between fossils and living plants and animals in the Old and the New Worlds. One theory contended that the continents did not actually move away from each other but that the ocean basins between them developed from sinking land, which at one time linked the continents. Since most geologists at that time thought the Earth was contracting, as the interior cooled and shrank, they argued, blocks of crust fell inward to fill the void spaces and in the process created the ocean basins.

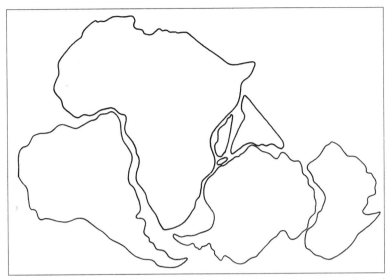

Figure 5 During the Paleozoic, the southern continents combined into Gondwana.

The belief that the Earth was shrinking in order to maintain its internal temperature was soon discarded with the discovery of radioactive decay near the turn of the 20th century. The decay of radioactive elements in the Earth's interior supplied the energy required to keep it hot. Many geologists then did a complete about-face and suggested that, instead of shrinking, the Earth was expanding to get rid of the excess heat.

EVIDENCE FOR DRIFT

The major problem with early theories dealing with the separation of the continents was that the phenomenon was thought to have commenced early in the Earth's history. Therefore, scientists were forced to devise complex theories to account for the similarities among plant and animal fossils that have been separated by oceans for eons. It seemed highly unlikely that so large a variety of species could have evolved along parallel lines in such diverse environments over such a lengthy period.

Figure 6   Lystrosaurus and fossil sites in Gondwana.

In the frigid cliffs of the Transantarctic Mountains of Antarctica, scientists discovered a fossilized jawbone and canine tooth belonging to the mammal-like reptile lystrosaurus (Figure 6). This unusual-looking animal, with large down-pointing tusks, lived around 160 million years ago. The only other known lystrosaurus fossils were found in China, India, and southern Africa. It seemed very unlikely that this freshwater reptile crossed the salty oceans that separated the southern continents. Instead, its discovery on the frozen wastes of Antarctica was hailed as definite proof for the existence of Gondwanaland (now called Gondwana).

In 1985, a small fossil opossum tooth that dated about 37 million years old was discovered in central Siberia. Opossums originated in North America during the late Cretaceous, about 85 million years ago. The animal could have taken a northern route through Asia, leading eventually to Australia, or the direct southern route through South America (Figure 7). The southern route theory is supported by a fossil of a South American marsupial found in Antarctica, which would have acted as a land bridge between the southern tip of South America and Australia, lending further support to the existence of Gondwana.

Further evidence of continental drift include the fossils of a 270-million-year-old reptile called mesosaurus (Figure 8), found in Brazil and South Africa. This 2-foot-long reptile lived in shallow, freshwater lakes, and it was highly unlikely that it swam across the thousands of miles that separated the two continents. Fossils of the late Paleozoic fern glossopteris were found in Australia and India but were conspicuously missing on the northern continents. This suggested that at one time there were two large landmasses: one located in the Southern Hemisphere, the other in the Northern Hemisphere. The continents were separated by a wide sea, and their breakup appeared to be relatively late in Earth's history.

Not only did plant and animal life on widely dispersed continents seem to have common ancestors, but the older rocks were remarkably similar as well. There were also matching rocks of several mountain ranges (Figure 9). The Cape Mountains in South Africa were very similar to the Sierra Mountains south of Buenos Aires in Argentina. Furthermore, there were matches between mountains in Canada, Scotland, and Norway. Not only were the rock strata the same type and age, but they were also laid down in exactly the same order.

The continents of Africa, South America, Australia, Asia, and Antarctica showed evi-

Figure 7   Possible routes for marsupials from North America to Australia.

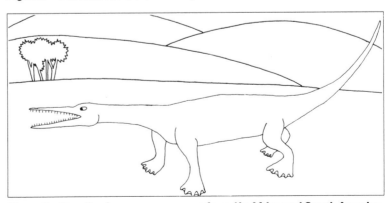

Figure 8   Fossils of mesosaurus were found in Africa and South America.

dence of contemporaneous glaciation in the late Paleozoic, around 270 million years ago as well. This is indicated by deposits of glacial till and grooves in the ancient rocks excavated by boulders embedded in slowly moving masses of ice. The lines of ice flow were away from the equator and toward the poles, which would not be the case if the continents were situated as they are today since glacial centers do not exist on the equator. Thus, the continents must have been joined so that the ice moved across a single landmass, radiating outward from a glacial center over the South Pole.

Strange out-of-place boulders called erratics, composed of rock types not found elsewhere on one continent, also matched up with rocks on the opposite continent. The glacial deposits were overlain by thick sequences of terrestrial deposits, which in turn were overlain by massive outpourings of basalt lava flows. Overlying these deposits were coal beds containing similar fossilized plant material.

Glacial deposits located in the equatorial areas suggested that in the past these regions were much colder. Coral reefs and coal deposits found in the North Polar regions indicated that a tropical climate once existed there. Morever, in the arctic regions, salt deposits indicated an ancient desert climate. Either the climate in the past changed dramatically, or the continents changed position with respect to the equator.

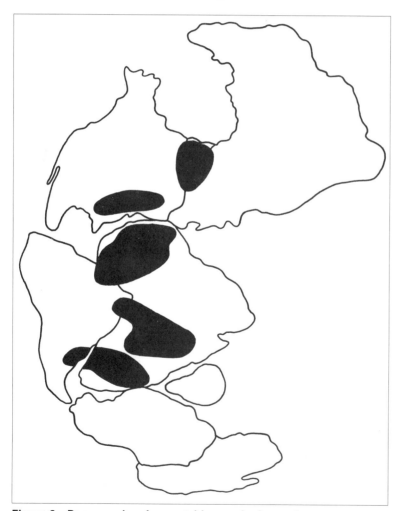

Figure 9   Pangaea showing matching geologic provinces.

CONTINENTAL MOVEMENTS

Early 20th-century geologists still held to the belief that narrow land bridges spanned the distances between continents. The idea was that the continents were always fixed and that land bridges rose up from the ocean floor to allow the migration of species from one continent to another. Later, the land bridges sank out of sight beneath the surface of the sea.

However, searching for evidence of land bridges by sampling the ocean floor failed to turn up even a trace of sunken land, and new theories were devised which also cast doubt on the existence of land bridges. An American geologist, Frank Taylor, was opposed to the idea that continents sank into the ocean floor, simply because they were lighter than the underlying basalts. In 1908, he suggested an alternative explanation for the formation of mountains based on continental movements. Taylor thought that two great landmasses located at the poles slowly crept toward the equator and their collision shoved up great blocks of crust into mountain ranges. He also described an undersea mountain range between South America and Africa, known today as the Mid-Atlantic Ridge. He believed that it was a line of rifting between the two continents. The ridge remained stationary, while the two continents crept away from it in opposite directions.

The German meteorologist and Arctic explorer Alfred Wegener was intrigued by the high degree of correspondence between the shapes of continental coastlines on either side of the Atlantic Ocean. He, too, argued that a land bridge was not possible because the continents stand higher than the seafloor for the simple reason that they are composed of light granitic rocks that float on the denser basaltic rocks of the upper mantle. It was inconceivable for lighter rocks to sink into heavier ones. To Wegener it seemed the most likely explanation was that the continents were once connected and subsequently drifted apart.

Wegener published his theory of continental drift in 1915. He believed that 200 million years ago all the lands constituted a single large continent, which he named Pangaea, meaning "all lands." The rest of the world was covered by an ocean he called Panthalassa, meaning "universal sea." When Pangaea broke up and the continents drifted apart, the Atlantic and Indian Oceans were formed. Wegener supported his hypothesis with an impressive collection of evidence, including the geometric fit of continental margins, matching mountain chains on opposite continents, corresponding rock successions, similar ancient climatic conditions, and identical species on continents now widely separated by ocean. The continents were like torn pieces of a newspaper—not only did the edges match when fitted together, but the printed words did as well.

Before Wegener introduced his theory, the formation of mountain ranges was poorly understood. Geologists assumed that mountains formed when the molten crust solidified and shriveled up. After making more extensive studies of mountain ranges, however, they were forced to conclude that the folding of rock layers was much too intense (Figure 10), requiring a considerably more rapid cooling and contraction than the Earth could

Figure 10   Folded Cambrian rocks on the south side of Scapegoat Mountain, Lewis and Clark County, Montana. Photo 131 by M. R. Mudge, courtesy USGS

possibly have withstood. Moreover, under these conditions, mountains would have been scattered evenly throughout the world instead of concentrated in a few long mountain belts.

According to Wegener, as the continents made their way across the ocean floor following their breakup, they encountered increasing resistance, which forced the leading edges to crumble, fold back, and thrust upward. Wegener pointed to the long sinuous Rocky and Andes Mountains in North and South America as classic examples of this type of folded mountain belt.

Wegener's continental drift theory drew fire from most contemporary geologists, who questioned whether the soft, light rocks of the continents could penetrate the hard, dense rocks of the ocean floor. They were opposed to the idea that the breakup of the continents occurred so late in geologic history. Geologists also insisted that erosion was responsible for the matching coastlines of South America and Africa, and that the close fit of the continents was merely a coincidence.

Wegener also failed to devise a satisfactory mechanism for the movement of the continents. He thought that the rotation of the Earth provided the necessary force. He contended that as the Earth spins on its axis, centrifugal force made the outer layers fly outward, pushing them away from the poles, producing a bulge at the equator. The continents would simply slide down the slope of the bulge under the force of gravity.

Other scientists countered Wegener with scientific data. Studies of deep earthquakes at the ocean-continent boundaries, especially around the Pacific, demonstrated the deep structure of the continents with their roots well embedded in the upper mantle. In 1909, the Yugoslavian seismologist Andrija Mohorovicic discovered the division between the mantle and the crust known as the Mohorovicic discontinuity, or simply Moho (Figure 10a). By studying seismic waves generated by earthquakes, seismologists were able to determine certain properties of the Earth's interior. Furthermore, calculations of the Earth's heat flow suggested that the continents were formed from the mantle beneath them. If drift had occurred, there would be uneven patterns of heat flow. This placed severe restrictions on continental drift, for the continents were much too thick and therefore anchored securely in place.

Because Wegener worked so hard to prove his theory, he tended to exaggerate and saw evidence where none existed. Using inaccurate land surveys in Greenland, he calculated its rate of drift at a phenomenal 40 yards a year, which would have it circle the Earth every million years. When Wegener died of a heart attack during an expedition to Greenland in 1930, the continental drift theory largely died with him.

The rejection of continental drift is a clear example of how old scientific theories become entrenched. Once a theory is accepted as fact by the scientific community, in this case, of land bridges, it becomes a solemn doctrine, and more effort is put into proving it than into disproving it. Geologists were caught in the middle of a scientific revolution in which hard evidence was simply ignored for fear that years of painstaking research would have to be thrown out. Like Charles Darwin's theory of evolution, if the continental drift theory was ever spoken of, it was held up to ridicule and contempt and considered a classic scientific blunder. Wegener was also considered an outsider and not part of the geologic fraternity. This might have worked to his advantage, however, for he was less reluctant to discard outmoded geologic thinking.

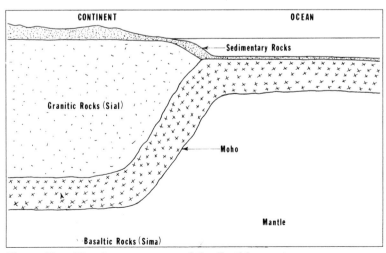

Figure 10a The deep structure of the Earth's crust.

PALEOMAGNETISM

Further evidence for continental drift was found in studies of the Earth's magnetic field. The magnetic field is believed to originate from electric currents in the core. These currents could be generated by several means, such as the chemical differences between the inner and outer core. Once these currents are initiated, they are amplified by the dynamo effect created by the core as it rotates, which acts like an electrical generator. Since the core is composed of iron and nickel, both good electrical conductors, the electric currents passing through it set up a weak magnetic field within the core. The motion of the core reinforces the current, which in turn generates a larger magnetic field.

Because the Earth rotates, the magnetic field tends to be aligned in one direction. Apparently the Earth's magnetic field remains stable for long periods. Then for unknown reasons, the electric currents fail and the magnetic field collapses. Eventually, the field is regenerated with opposite polarity with north where south used to be and vice versa. When iron-rich basalt lava cools, the magnetic fields of its iron molecules are aligned with the Earth's magnetic field like miniature bar magnets. In essence, they are like fossilized compasses, pointing in whatever direction North happens to be at the time of deposition.

When sensitive magnetic recording instruments called magnetometers were first taken into the field in the 1950s, scientists in England were very much perplexed by their findings. Rocks formed 200 million years ago showed a mag-

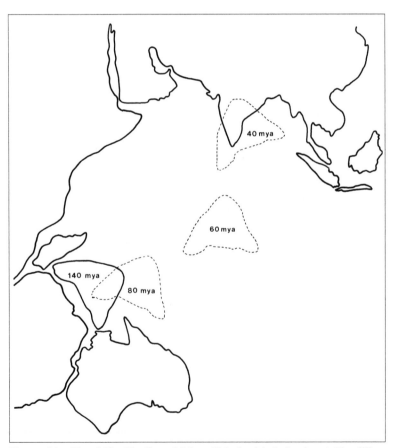

Figure 11   The drift of India, which collided with Asia about 40 million years ago.

netic inclination (the downward pointing of the needle of a vertically held compass) of 30 degrees north, whereas England's present inclination is 65 degrees north. The inclination is almost 0 degrees at the equator and 90 degrees at the poles. The only conclusion that could be drawn from the data was that England must have once been farther south.

To further test this astounding discovery, the scientists took their instruments to India's Deccan Plateau. Rocks dating 150 million years old showed a magnetic inclination of 64 degrees south. Rocks that were 50 million years old indicated an inclination of only 26 degrees south. But what was even more astonishing was that rocks dating 25 million years old completely reversed their magnetic inclination and read 17 degrees north. Apparently, India at one time was in the southern hemisphere and had crossed over the equator into the northern hemisphere to its present location as part of Asia (Figure 11).

Skeptics pointed out, however, that the same phenomenon could occur simply by the shifting of the Earth's magnetic poles. Such polar wandering would have altered the direction of the Earth's magnetic field, and records of these changes would be permanently locked up in the rocks. Therefore, rocks formed at different times in England and India could have been imprinted with different inclinations without moving an inch. Thus, the evidence for continental drift was used to prove that it was actually the *North Pole* that had wandered some 13,000 miles over the last billion years from western North America, across the northern Pacific Ocean and northern Asia, finally coming to rest at its present location in the Arctic Ocean.

When similar magnetometer measurements were taken in North America, however, the results came as a com-

Figure 12   Polar wandering curves for North America (top curve) and Europe (bottom curve) veer apart due to continental drift.

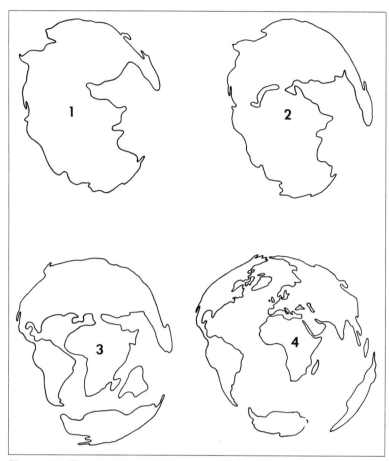

Figure 13 The breakup and dispersal of the continents: (1) Pangaea 200 mya; (2) opening of the North Atlantic Ocean 160 mya; (3) the continents at the height of the dinosaur age 80 mya; (4) the present.

plete surprise. Although the polar paths derived from data on Eurasia and North America both were much the same shape and had a common point of origin at the North Pole, the curves gradually veered away from each other (Figure 12.) Only by hypothetically joining the continents together did the two curves overlap. Thus, in their efforts to disprove continental drift, scientists inadvertently provided the strongest evidence in its favor.

LITHOSPHERIC PLATES

Only after overwhelming geologic and geophysical evidence collected from the ocean floor, and supporting the theory of continental drift did geologists finally abandon the archaic thinking of the past century. By the late 1960s, most geologists in the Northern Hemisphere, who had fought hard against the theory, joined their southern colleagues, who for some time were convinced of the accuracy of the continental drift model.

The generally accepted model described in detail in the next chapter, is that the present continents were sutured together into the supercontinent Pangaea during the Permian and early Triassic. In the late Triassic and early Jurassic, Pangaea began to rift apart along a fracture zone now represented by the Mid-Atlantic Ridge. The breakup finally became complete sometime during the early Cretaceous, and since then all continents bordering on the Atlantic have been drifting away from each other (Figure 13).

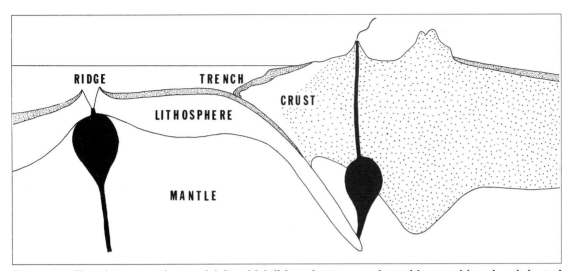

Figure 14   The plate tectonics model, in which lithosphere created at midocean ridges is subducted into mantle at deep-sea trenches and is responsible for all geologic activity.

Despite the mounting evidence supporting continental drift, skeptics continued to doubt the breakup and drift of the continents and questioned whether the currents in the Earth's interior were powerful enough to propel the continents around. Even some supporters of the continental drift theory thought this energy source might not be sufficient and suggested that additional mechanisms, such as gravity, were needed to move the continents. Furthermore, if Earth processes were essentially uniform throughout geologic time, why did the breakup of the continents happen so late in the Earth's history? And were there earlier episodes of continental collision and breakup?

Regardless of these objections, the overwhelming majority of Earth's scientists accepted continental drift as a scientific certainty. So convincing was the evidence that, by the late 1960s, it gave rise to an entirely new way of looking at the Earth, called the theory of plate tectonics (tectonics is any process by which the Earth's surface is shaped). The plate tectonics theory (Figure 14) was first publicized by the British geophysicists Dan McKenzie and Robert Parker in 1967. It incorporated the processes of seafloor spreading and continental drift into a single comprehensive model. Therefore, all aspects of the Earth's history and structure could be unified by the revolutionary concept of movable plates. The boundaries of the plates were marked by well-defined earthquake zones, and analysis of earthquakes around the Pacific Ocean revealed a consistent direction of crustal movement.

The theory was also developed independently by the American geophysicist Jason Morgan at Princeton University. He regarded the Earth's outer shell as neither rigid nor fixed but made up of several movable plates about 60 miles thick. Each plate was composed of the lithosphere, the solid portion of the upper mantle, and the overlying continental or oceanic crust. The plate boundaries were the midocean ridges, where new oceanic crust is created as the plates are pulled apart; the transform faults, where the plates slide past each other, often wrenching the ocean floor in the process; and the deep-sea trenches, where the plates are subducted or absorbed into the mantel and destroyed. The plates ride on the asthenosphere, the semimolten layer of the upper mantle, and carry the continents along with them like rafts of rock riding on a sea of molten magma.

2

HISTORIC TECTONICS:
AN OVERVIEW

Plate tectonics has been shaping the Earth almost since its beginning. Continents were set adrift practically since they were first
formed some 4 billion years ago. This is manifested by ancient
granites discovered in the Northwest Territories of Canada, which
suggest that the formation of the crust was well under way by this
time because the presence of granites is an indication of crustal
generation. The continental crust was only about one-tenth its present size and contained slivers of granite that drifted freely over the
Earth's watery surface.

As time went on, these slices of crust began to slow their erratic
wanderings and combined into larger landmasses. Continuous bumps
and grinds built up the crust until by the beginning of the Proterozoic
Eon 2.5 billion years ago the landmass occupied up to one quarter of the
Earth's surface area. It was also during this time that plate tectonics
began to operate extensively, and the Earth as we know it began to take
shape.

TABLE 1   THE GEOLOGIC TIME SCALE

Era	Period	Epoch	Age in Millions of Years	First Life Forms
		Holocene	0.01	
	Quaternary			
		Pleistocene	2	Man
Cenozoic		Pliocene	7	Mastodons
		Miocene	26	Saber-toothed tigers
	Tertiary	Oligocene	37	
		Eocene	54	Whales
		Paleocene	65	Horses Alligators
	Cretaceous		135	
				Birds
Mesozoic	Jurassic		190	Mammals
				Dinosaurs
	Triassic		240	
	Permian		280	Reptiles
		Pennsylvanian	310	
				Trees
	Carboniferous			
Paleozoic		Mississippian	345	Amphibians Insects
	Devonian		400	Sharks
	Silurian		435	Land plants
	Ordovician		500	Fish
	Cambrian		570	Sea plants Shelled animals
			700	Invertebrates
Proterozoic			2500	Metazoans
			3500	Earliest life
Archean			4000	Oldest rocks
			4600	Meteorites

ARCHEAN TECTONICS

The first two billion years of Earth history is known as the Archean Eon. It was a time when the Earth's interior was much hotter than it is today, probably too hot for plate tectonics to operate effectively as there was more vertical bubbling than horizontal sliding. Basalt, which is a dark, dense volcanic rock, formed most of the early crust, both on the continents and under the oceans. The crust was thin and highly unstable. When tectonic activity melted and remelted the basaltic crust, granites were formed. Eventually, the basaltic crust became embedded with scattered blocks of granite called *rockbergs.*

The formative Earth was subjected to massive volcanism and meteorite bombardment that repeatedly destroyed the crust. This is why the first 700 million years of Archean time is missing from the geologic record. Heavy turbulence in the mantle, with a heat flow three times greater than it is today, resulted in violent agitation on the surface. This produced a sea of molten and semimolten rock broken up by giant fissures, from which fountains of lava spewed skyward.

About 4 billion years ago, during the height of the great meteorite bombardment, a massive asteroid landed in what is now central Ontario, Canada (Figure 15). The impact created a crater up to 900 miles wide, which might have triggered the formation of the continental crust by melting vast amounts of basalt, converting it into granitic rock. Few rocks on Earth date beyond 3.7 billion years, which means that little continental crust was generated until after that time or was recycled before then. Slices of granitic crust combined into stable bodies of basement rock, upon which all other rocks were deposited. The basement rocks formed the nuclei of the continents and are presently exposed in broad, low-lying, domelike structures called shields.

Archean greenstone belts are dispersed

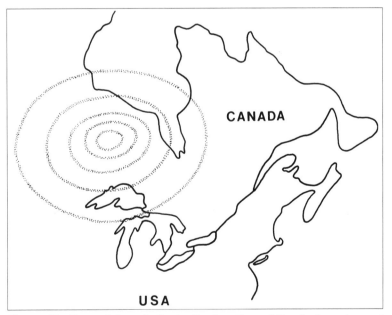

Figure 15 Dotted rings indicate the location of the Archean impact structure in central Ontario.

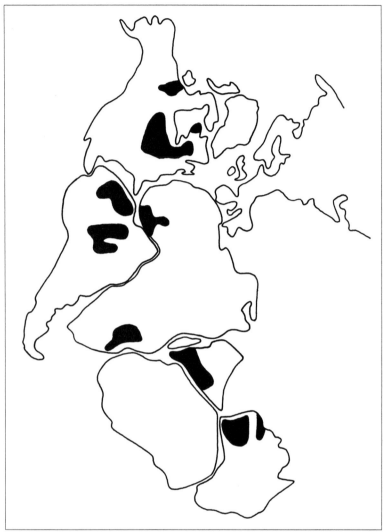

Figure 16   Archean greenstone belts, indicated by dark areas, are the earliest evidence of plate tectonics.

among and around the shields, and composed of a jumble of metamorphosed (recrystallized) lava flows and sediments, possibly from chains of volcanic islands called island arcs, caught between colliding continents. Their green color is derived from the mineral chlorite, a greenish form of mica. Also caught in the greenstone belts were ophiolites, slices of ocean floor shoved up on the continents by drifting plates, which are 3.6 billion years old. Ophiolites provide the best evidence for ancient plate motions. They are vertical cross sections of oceanic crust that were peeled off during plate collisions and plastered onto continents. This resulted in a linear formation of greenish volcanic rocks along with light-colored masses of granite and gneiss, which are common igneous and metamorphic rocks.

Pillow lavas, which are tubular bodies of basalt extruded undersea, are also found in the greenstone belts, signifying that the volcanic eruptions took place on the ocean floors. Greenstone belts are found in all parts of the world and occupy the ancient cores of the continents (Figure 16). They are of particular interest to geologists, not only as evidence for Archean plate tectonics but also because they hold most of the world's gold deposits.

Only three sites, located in Canada, Australia, and Africa, contain rocks exposed on the surface during the Archean that have remained relatively unchanged throughout geologic time. In the Barberton Greenstone Belt of South Africa lies a thick, widespread bed of silicate spherules, which are small glassy beads believed to have originated from the melt of a large meteorite impact between 3.5 and 3.2 billion years ago. This was a time when large impacts were quite numerous and played a more prominent role in shaping the Earth's surface than they do today.

The oldest evidence of life on Earth is microfossils, the remains of ancient microorganisms, and stromatolites, the layered structures formed by the accretion of fine sediment grains by colonies of primitive blue-green algae. The earliest stromatolites were found in sedimentary rocks of the Warrawoona group in Western Australia that are 3.5 billion years old. Associated with these rocks are cherts containing microfilaments, which are small threadlike structures of possible bacterial origin. Similar cherts with mircrofossils of primitive bacteria were found in 3.3-billion-year-old rocks from eastern Transvaal in South Africa.

The abundance of chert in deposits older than 2.5 billion years indicates that most of the crust was deeply submerged during this time because it precipitates in silica-rich seawater. The seas contained much more dissolved silica, which leached out of the volcanic rock pouring unto the ocean floor. Modern ocean water is deficient in silica because organisms like sponges and diatoms extract it to build their skeletons. Massive deposits of diatomaceous earth in many parts of the world are a tribute to the great success of these organisms.

CRATONS

Cratons contain the oldest rocks found on Earth, dating back to 4 billion years. These rocks were composed of highly altered granite and metamorphosed marine sediments and lava flows. They originated from intrusions of magma into the primitive ocean crust. This allowed the magma to slowly cool and separate into a light component, which rose toward the surface, and a heavy component, which settled to the bottom of the magma chamber.

Some of the magma also seeped through the crust, where it poured out as lava on the ocean floor. Successive intrusions and extrusions of magma built up new crust until it finally broke the surface of a global sea. Unlike the volcanic islands, which existed only for a short time, these thin slivers of land became a permanent part of the landscape, since the cratons were lighter and more buoyant than the ocean crust. If a craton was forced into the mantle under the pull of gravity, it would bob up again like a cork on the end of a fishing line.

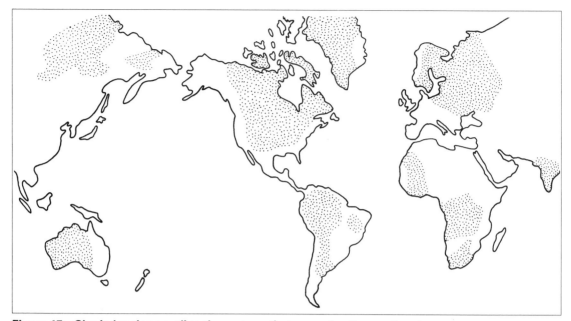

Figure 17   Shaded regions outline the cratons that comprise the continents.

The cratons, which numbered in the dozens, slowly built up and ranged from about a fifth the size of present-day North America to smaller than the state of Texas. The cratons were also very mobile and moved about freely, becoming independent, freewheeling minicontinents that periodically collided with and rebounded off each other. All the cratons eventually coalesced into a single large landmass several thousands of miles wide. The point at which the cratons collided with one another became mountain ranges, and the sutures joining the landmasses are still visible today as cores of ancient mountains over 2 billion years old. The collisions caused a slight crumpling at the leading edges of the cratons, forming small parallel mountain ranges perhaps only a few hundred feet high.

Volcanoes were highly active on the cratons, and lava and ash continued to build them upward and outward. New crustal material was also added to the interior of the cratons from magmatic intrusions composed of molten crustal rocks that were recycled through the Earth's upper mantle. This effectively cooled the mantle, causing the cratons to slow down their erratic wanderings. Some of the original cratons formed within the first 1.5 billion years of the Earth's existence, and totaled about 10 percent of the present landmass. The average rate of continent growth since the Earth was formed was perhaps as much as one cubic mile a year. The constant rifting and patching of the interior along with sediments deposited along the continental margins eventually built the continents outward until their area was nearly equal to the total area of all the present-day continents.

PROTEROZOIC TECTONICS

The Proterozoic Eon, from about 2.5 billion years ago to about 600 million years ago, witnessed a dramatic change in the Earth as it matured from a tumultuous adolescence to a more tranquil adulthood. At the beginning of the eon, nearly three quarters of the present landmass was in existence. It was composed of Archean cratons, pieces of granitic crust welded together to form the cores of the continents (Figure 17). Many of these cratons throughout the world were assembled at about the same time.

As the cratons grew more sluggish, they developed a greater tendency to stick together after collision. The North American continent comprises seven cratons that assembled around 2 billion years ago (Figure 18). Successive continental collisions added new crust to the growing proto–North American continent. A major portion of the continental crust underlying the central part of the United States formed in one great surge of crustal generation around 1.8 billion years ago that has been unequaled since. This was possibly the most energetic period of tectonic activity and crustal generation in Earth history. The assembled North American continent was stable enough to resist another billion years of jostling and rifting and continued to grow by plastering bits and pieces of continents and island arcs to its edges.

About 1.8 billion years ago, a large meteorite slammed into the continent in what is now Ontario, Canada, with enough energy to disturb crustal rocks for upward of 100 miles or more. The

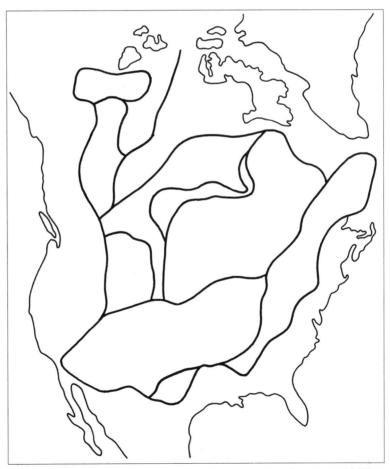

Figure 18   **The cratons of North America came together some 2 billion years ago.**

impact melted vast quantities of basalt and granite. Metals separated out of the molten rocks to form the world's largest and richest nickel ore deposit, known as the Sudbury Igneous Complex. It is also one of the world's oldest astroblems, which are ancient, eroded impact structures.

The presence of large amounts of volcanic rock near the eastern edge of North America implies that the continent was once the core of a larger supercontinent. The central portion of the supercontinent was far removed from the cooling effects of subducting plates, where the Earth's crust sinks into the mantle. As a result, the interior of the supercontinent heated up and erupted with volcanism. The warm, weakened crust consequently broke into possibly four or five major continents between 625 and 550 million years ago along these lines of volcanic activity.

Thick deposits of Proterozoic sediments were derived from Archean granites. Nearly 20,000 feet of these sediments lie in the Unita Range of Utah (Figure 19), and the Montana Proterozoic belt system contains sediments over 11 miles thick. Terrestrial red beds, whose sediment grains are cemented by iron oxide, which gives the deposits their red color, appeared about 1 billion years ago, signifying that the atmosphere had significant amounts of oxygen by this time.

Figure 19   Red Castle Peak and Lower Red Castle Lake, Unita Mountains, Summit County, Utah. Photo 145 by W. R. Hansen, courtesy USGS

Around 670 million years ago, thick ice sheets spread over much of the landmass during perhaps the greatest period of glaciation the Earth has ever known. At this time, all continents were assembled into a supercontinent that might have wandered over one of the poles. When the ice age ended, life took off in all directions, resulting in many unique and bizarre creatures (Figure 20), as evidenced by fossil impressions found in the Ediacara formation in South Australia. Toward the end of the Proterozoic, the supercontinent, now located near the equator, broke up, and the increased habitat area resulted in the greatest explosion of new species the world has ever known.

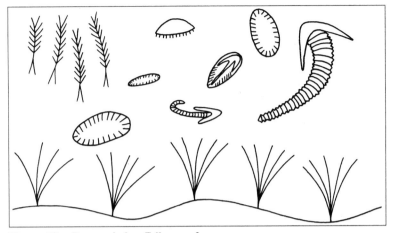

Figure 20 Precambrian Ediocara fauna.

PALEOZOIC TECTONICS

The Paleozoic era began about 570 million years ago and ended about 240 million years ago. The first half of the era was tectonically quiet, with little mountain building and volcanic activity. The land was divided into two megacontinents. The northern landmass was called Laurasia and included what is now North America, Greenland, Europe, and Asia. The southern landmass was called Gondwana and included what is now Africa, South America, Australia, Antarctica, and India. The subcontinent of India later broke away from Gondwana and drifted into southern Asia. The collision uplifted the Himalaya Mountains and the broad Tibetan Plateau, most of which rises over 3 miles above sea level.

The two great landmasses were separated by a large body of water called the Tethys Sea, which held thick deposits of sediments washed off the continents. The continents were lowered by erosion, and shallow seas flooded inland, covering more than half the present land area. The weight of the sediments formed a deep depression in the ocean crust called a geosyncline. It was later uplifted into great mountain belts surrounding the Mediterranean Sea, when Africa slammed into Europe.

Continental movements are thought to be responsible for a period of glaciation that occurred during the late Ordovician around 440 million

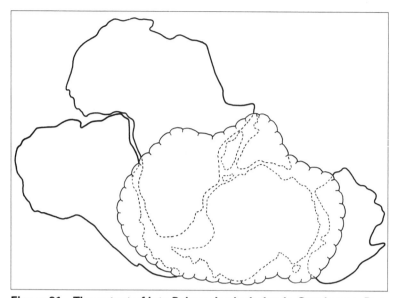

Figure 21   **The extent of late Paleozoic glaciation in Gondwana. Dotted lines indicate how the present continents were affected.**

years ago. The study of magnetic orientations in rocks from many parts of the world indicates the positions of the continents relative to the magnetic poles at various times in geologic history. The paleomagnetic studies in Africa were very curious, however, for the northern part of the continent was placed directly over the South Pole during the Ordovician, which led to widespread glaciation.

Evidence for this period of glaciation came from an unexpected source. In the middle of the Sahara Desert, geologists exploring for oil stumbled upon a series of giant grooves that appeared to be cut into the underlying strata by glaciers. The scars were created by rocks embedded at the base of glaciers as they scraped the bedrock. Further evidence that the Sahara Desert had once been covered by thick sheets of ice included erratic boulders carried long distances by the glaciers, and sinuous sand deposits from glacial outwash streams.

Another episode of glaciation occurred during the late Paleozoic, when Gondwana passed into the South Polar region and glacial centers expanded in all directions. Ice sheets covered large portions of east central South America, South Africa, India, Australia, and Antarctica (Figure 21). During the early part of the glaciation, the maximum glacial effects occurred in South America and South Africa. Later, the chief glacial centers switched to Australia and Antarctica, providing strong evidence that the southern continents had wandered locked together over the South Pole.

In Australia, marine sediments were found interbedded with glacial deposits, and tillites, composed of glacially deposited boulders and clay, were separated by seams of coal, indicating that periods of glaciation were punctuated by warm interglacial spells, when extensive forests grew. The Karroo Series in South Africa is composed of a sequence of late Paleozoic tillites and coal beds, reaching a total thickness of 20,000 feet. In between layers of coal were fossil leaves of an extinct fern called glossopteris (Figure 22). Because this plant is found only on the southern continents, it is among the best evidence that they were once joined together as Gondwana.

When the Paleozoic came to a close, Gondwana and Laurasia converged into the crescent-shaped supercontinent Pangaea (Figure 23). The supercontinent extended practically from pole to pole. The rest of the planet was covered by a single great ocean called the Panthalassa. The continental collisions crumpled the crust and pushed up huge masses of rocks into several mountain belts throughout many parts of the world. Volcanic eruptions were also prevalent due to frequent continental collisions.

The sediments in the Tethys Sea separating Gondwana and Laurasia were squeezed and uplifted into various mountain belts, including the Ouachitas and Appalachians of North America and the ancestral Hercynian Mountains of southern Europe. As the continents rose higher and the ocean basins dropped lower, the land became dryer and the

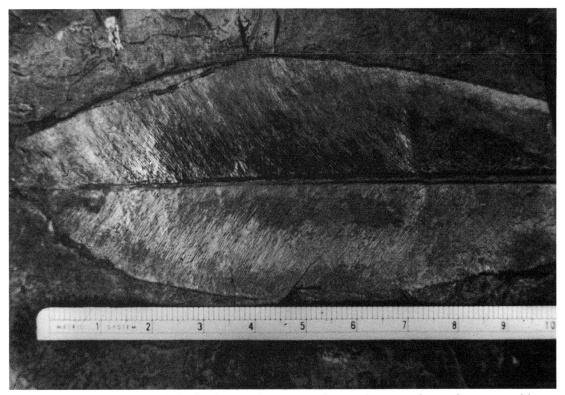

Figure 22 Fossil glossopteris leaf, whose existence on the southern continents is strong evidence for Gondwana. Photo 23 by D. L. Schmidt, courtesy USGS

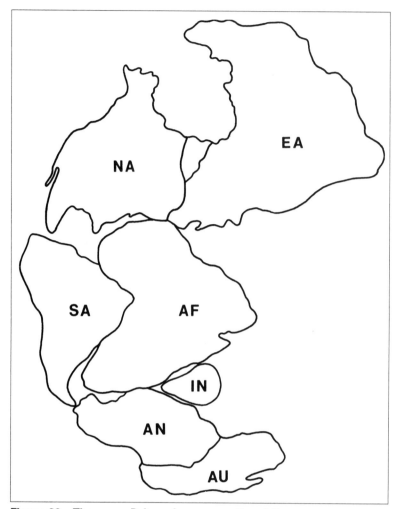

Figure 23   The upper Paleozoic supercontinent Pangaea.

climate grew colder, especially for the southernmost lands, which were covered with glacial ice.

MESOZOIC TECTONICS

At the beginning of the Mesozoic era, which spans from about 240 million to 65 million years ago, the great glaciers of the previous ice age melted and the seas began to warm. The once towering mountain ranges of North America and Europe were toppled by erosion. Reef building was intense in the Tethys Sea, and thick deposits of limestone and dolomite were laid down by lime-secreting organisms. These beds were later uplifted to form the dolomite and limestone Alps.

Terrestrial red beds covered much of North America, including the Colorado Plateau and a region from Nova Scotia to South Carolina. They were also common in Europe. Such wide occurrences of red sediments were probably due to massive accumulations of iron supplied by one of the most intense periods of igneous activity the world has ever known. There were huge lava flows and granitic intrusions in Siberia, and extensive lava flows covered South America, Africa, and Antarctica as well. Southern Brazil was paved over with some 750,000 square miles of basalt, constituting the largest lava field in the world.

Great floods of basalt, in places 2000 feet or more thick, covered large parts of Brazil and Argentina when the South American plate overrode the

Pacific plate, causing it to subduct and melt to feed magma chambers underlying active volcanoes. Basalt flows also occurred from Alaska to California. In addition, there were massive granitic intrusions, such as the huge Sierra Nevada batholith in California.

These tremendous outpourings of basalt reflected one of the greatest crustal movements in the history of the planet. The continents probably traveled much faster than they do today, because of more vigorous plate motions, resulting in greater volcanic activity. Around the close of the Triassic period, North and South America separated; India, nestled between Africa and Antarctica, began to separate from the two continents and move northward; and a great rift began to separate the North American continent from Eurasia. The rift was later breached and flooded with seawater, forming the infant North Atlantic Ocean.

During the Jurassic period which followed, South America began to separate from Africa like a zipper opening up from south to north. India was set fully adrift and headed toward southern Asia. Antarctica, still attached to Australia, swung away from Africa toward the southeast, forming the proto–Indian Ocean (Figure 24). North America drifted westward as the North Atlantic continued to widen at the expense of the Pacific. Because of the effects of seafloor spreading, which creates new oceanic crust, in the Atlantic, and plate subduction, which destroys old oceanic crust, in the Pacific, the crust of the Pacific Basin is no older than middle Jurassic in age.

Much of western North America was assembled from oceanic island arcs and other crustal debris that were skimmed off the Pacific plate as the North American plate continued heading westward. Northern California is a jumble of crust assembled only a few hundred million years ago. A nearly complete slice of ocean crust, the type that is shoved up on the continents by drifting plates, sits in the middle of Wyoming.

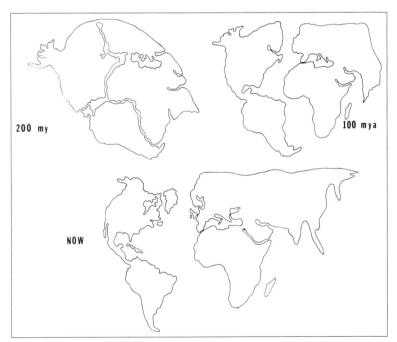

200 my

100 mya

NOW

Figure 24 **The breakup and drift of the continents to their present localities.**

The Tethys Sea provided a wide gulf between the northern and southern landmasses and continued to fill with thick layers of sediment. The Western Interior Cretaceous Sea flowed into the west-central portions of North America (Figure 25), and accumulations of marine sediments eroded from the cordilleran highlands to the west were deposited on the terrestrial red beds of the Colorado Plateau.

During the Cretaceous period, great deposits of limestone and chalk were laid down in the interior seas of Europe and Asia (*creta* is Latin for chalk). Seas also invaded North and South America, Africa, and Australia. The oceans of the Cretaceous were interconnected in the equatorial regions by the Tethys and Central American seaways, providing a unique circumglobal current system (Figure 26). The continents were flatter, mountain ranges were lower, and sea levels were higher.

The interior seas that flooded North America were filled with thick deposits of sediment that are presently exposed as impressive cliffs in the western United States (Figure 27). The Appalachians, which were upraised by a continental collision between North America and northwest Africa near the beginning of the Triassic, were eroded to stumps by the Cretaceous. There was a high degree of geologic activity around the rim of the Pacific Basin, and practically all the mountain ranges facing the Pacific and the island arcs along its perimeter developed during this time.

When the Cretaceous came to a close, North America and Europe were no longer in contact except for a land bridge that spanned Greenland, the world's largest island. Greenland separated from North America and Norway during the early part of the Cenozoic era. The Bering strait between Alaska and Asia narrowed, creating a nearly

Figure 25   The Western Cretaceous Interior Sea in North America.

landlocked Arctic Ocean. The South Atlantic continued to widen at a rate of over an inch per year. South America and Africa were now separated by over 1500 miles of ocean. Africa moved northward, leaving Antarctica, still joined to Australia, far behind. As the African continent approached southern Europe, it began to pinch off the Tethys Sea, which was caught in the middle. Meanwhile, India continued to narrow the gap between it and southern Asia.

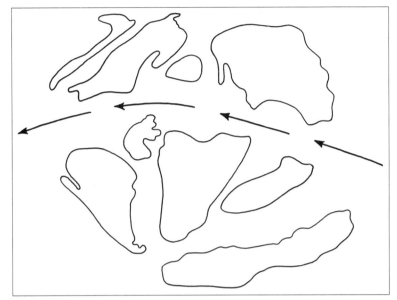

Figure 26 Mid-Cretaceous circumglobal current in the Tethys Sea that separated the northern and southern continents.

TABLE 2 THE DRIFTING OF THE CONTINENTS

Age (millions of years)		Gondwana	Laurasia
Quaternary	3		Opening of Gulf of California
Pliocene	11	Begin spreading near Galapagos Islands	Change spreading directions in eastern Pacific
		Opening of the Gulf of Aden	
			Birth of Iceland
Miocene	25		
		Opening of Red Sea	
Oligocene	40		
		Collision of India with Eurasia	Begin spreading in Arctic Basin

Age (millions of years)		Gondwana	Laurasia
Eocene	60	Separation of Australia from Antarctica	Separation of Green-land from Norway
Paleocene	65		
		Separation of New Zea-land from Antarctica	Opening of the Labra-dor Sea
			Opening of the Bay of Biscay
		Separation of Africa from Madagascar and South America	Major rifting of North America from Eurasia
Cretaceous	135		
		Separation of Africa from India, Australia, New Zealand, and Antarctica	
Jurassic	180		
			Begin separation of North America from Africa
Triassic	230		
Permian	280		

CENOZOIC TECTONICS

The Cenozoic era, from about 65 million years ago to the present, is best known for its intense mountain building. Highly active tectonic forces established much of the terrain features found on Earth today. Volcanic activity was extensive, and great outpourings of basalt covered Washington, Oregon, and Idaho, creating the Columbia River Plateau (Figure 28). Massive floods of lava poured onto an area of about 200,000 square miles and in places reached 10,000 feet thick. In only a matter of days, volcanic eruptions spewed forth batches of basalt as large as 1200 cubic miles, forming lava lakes up to 450 miles across. There was extensive volcanism in the Colorado Plateau and Sierra Madre region in Northwest Mexico as well.

About 50 million years ago, a huge submerged plateau in the Indian Ocean separated into two platforms that now sit about 1200 miles apart. The plateau rose from the ocean floor more than 90 million years ago, when a series of volcanic eruptions poured out vast amounts of molten basalt onto the Antarctic plate. During the next several million years, a long rift sliced through the plate and cut off its northern section, which latched onto

Figure 27   Sandstone cliffs of the Windgate formation in Colorado National Monument, Mesa County, Colorado. Photo 528 by J. R. Stacy, courtesy USGS

Figure 28 Palouse Falls in Columbia River basalt, Franklin-Whitman Counties, Washington. Photo 25 by F. O. Jones, courtesy USGS

the Indian plate and started on a journey northward. Meanwhile, the southern half of the plateau currently lies off the west coast of Australia. The other half is the world's largest submerged plateau and sits north of Antarctica.

The Rocky Mountains, which extend from Mexico to Canada, were pushed up during the Laramide orogeny (mountain building episode), which occurred from about 80 million to about 40 million years ago. During the Miocene, a large part of western North America was uplifted, and the entire Rocky Mountain region was raised about a mile above sea level. Great blocks of granite soared high above the surrounding terrain, while to the west in the Basin and Range Province, the crust was pulled apart and dropped in some places below sea level.

Around 30 million years ago, the North American continent began to approach the East Pacific Rise spreading center, which is the counterpart of the Mid-Atlantic Ridge. The first portion of the continent to override the axis of seafloor spreading was the coast of southern California and northwest Mexico. As the rift system and the subduction zone converged, the intervening oceanic plate was consumed in a deep trench. The sediments in the trench were caught in the big squeeze and heaved up to form the coastal ranges of California. At about the same time, Baja California was ripped from the mainland to form the

Gulf of California. Under similar circumstances, Arabia split off from Africa to form the Red Sea.

In northwest United States and British Columbia, the northern part of the East Pacific Rise was consumed in a subduction zone located beneath the continent. As the 50-mile-thick crustal plate was forced down into the mantle, the heat melted parts of the descending plate and the adjacent lithospheric plate, forming pockets of magma. The magma melted its way to the surface and formed the volcanoes of the Cascade Range. These erupted one after another in a great profusion, with Mount St. Helens among the most active of them all (Figure 29).

In South America, the mountainous spine of the Andes running along the western edge of the continent continued to rise throughout much of the Cenozoic due to the subduction of the Nazca plate underneath the South American plate. South America was temporarily connected to Antarctica by a narrow, curved land bridge. When Antarctica and Australia broke away from South America and moved eastward, they themselves separated. Antarctica moved toward the South Pole, while Australia continued in a northeasterly direction.

Near the end of the Eocene epoch, about 37 million years ago, global temperatures dropped significantly. Antarctica wandered over the South Pole and acquired a thick blanket of ice (Figure 30). Glaciers also grew for the first time in the highest ramparts of the Rocky Mountains. At times, Alaska connected with east Siberia at the Bering Strait, closing off the Arctic Basin from warm ocean currents, resulting in the accumulation of pack ice. A permanent ice cap did not develop over the North Pole, however, until about 4 million years ago, at which time Greenland acquired its first major ice sheet.

Between 3 and 4 million years ago, the Panama Isthmus, which separated North and South America was uplifted due to colliding

Figure 29   The May 18, 1980, eruption of Mount St. Helens, Skamania County, Washington. Photo courtesy USGS

Figure 30   A volcano practically buried in ice is evidence that tectonics were once active on Antarctica.
Photo 536 by W. B. Hamilton, courtesy USGS

oceanic plates. This halted the flow of cold water currents from the Atlantic into the Pacific, which together with the closing off of the Arctic Ocean from warm Pacific currents might have initiated the Pleistocene glacial epoch. Never before in Earth history had there been two permanent polar ice caps, which also suggests that the planet has been steadily cooling since the Cretaceous period, when the dinosaurs held dominion over the world.

TABLE 3   HISTORY OF THE DEEP CIRCULATION IN THE OCEAN

Age (mya)	Event
50	The ocean could flow freely around the world at the equator. Rather uniform climate and warm ocean even near the poles. Deep water in the ocean is much warmer than it is today. Only alpine glaciers on Antarctica.
35–40	The equatorial seaway begins to close. There is a sharp cooling of the surface and of the deep water in the south. The Antarctic glaciers reach the sea with glacial debris in the sea. The seaway between Australia and Antarctica opens. Cooler bottom water flows north and flushes the ocean. The snow limit drops sharply.
25–35	A stable situation exists with possible partial circulation around Antarctica. The equatorial circulation is interrupted between the Mediterranean Sea and the Far East.
25	The Drake Passage between South America and Antarctica begins to open.
15	The Drake Passage is open; the circum-Antarctic current is formed. Major sea ice forms around Antarctica, which is glaciated, making it the first major glaciation of the Modern Ice Age. The Antarctic bottom water forms. The snow limit rises.
3–5	Arctic glaciation begins.
2	An ice age overwhelms the Northern Hemisphere.

3

CONVECTION CURRENTS

The collision and separation of continents appears to be controlled by convection currents in the Earth's mantle. Convection is the motion within a fluid medium that results from a difference in temperature from top to bottom (Figure 31). Fluid rocks in the mantle receive heat from the core, ascend, dissipate heat to the lithosphere, cool, and descend to the core again to pick up more heat. The cycling of heat within the mantle is the main driving force behind all tectonics activity and, for that matter, all other activities taking place on the Earth's surface.

Rapid mantle convection leads to the breakup of supercontinents, which compresses the ocean basins, causing a rise in sea level and a transgression of the seas onto the land. There is also an increase in volcanism, and vast amounts of lava flood onto the continental crust during the early stages of many rifts. The rise in volcanism also increases the carbon dioxide content of the atmosphere, resulting in a strong greenhouse effect that promotes warm conditions worldwide.

The second phase of the continent cycle is a time of low mantle convection, resulting in the assembly of continents into a supercontinent. This process widens the ocean basins, causing a drop in sea level and a regression of the seas from the land. Furthermore, there is a decrease in volcanism and a reduction of atmospheric carbon dioxide, resulting in the development of an icehouse effect that leads to colder global temperatures.

HEAT FLOW

The formation of molten rock and the rise of magma to the surface results from the exchange of heat within the Earth's interior. Therefore, volcanism is linked to the Earth's temperature patterns. The Earth is continuously losing heat energy from the interior to the surface through its outer shell, or lithosphere, at a steady rate. About 70 percent of the heat flow results from seafloor spreading (described later in Chapter 5), and most of the rest is due to volcanism in the Earth's subduction zones (Chapter 6). Volcanic eruptions, however, only represent highly localized and spectacular releases of this energy (Figure 32).

The total heat loss does little toward heating the surface of the planet, where the vast majority of the surface

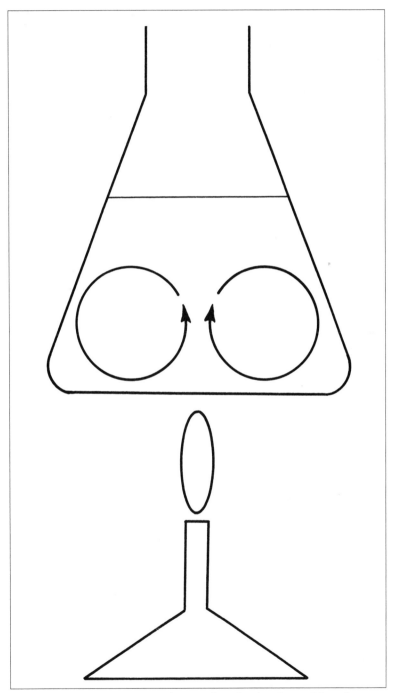

Figure 31 The movement of water in a beaker heated by a bunsen burner demonstrates the principle of convection.

Figure 32 The second eruption of Mount St. Helens, Skamania County, Washington, on July 22, 1980. Photo by Jim Vallance, courtesy USGS

heat comes from the sun and is thousands of times greater. Heat flow cannot be measured directly and depends on the temperature gradient and thermal conductivity of the rocks. As a rule, rocks are good thermal insulators, and continental crust is a better insulator than oceanic crust. If this were not so, the interior heat would have long since escaped into space, the Earth's interior would have cooled and solidified, and the continents would have stopped dead in their tracks.

Over the past several years, scientists have been taking the Earth's temperature. Thousands of heat flow measurements have been logged from around the world on land as well as at sea. On the ocean floor, a long, hollow cylinder is plunged into the soft sediments. Inside the buried cylinder, the temperature of the sediments is measured at intervals along its vertical length with fixed electrical thermometers. On land, exploratory boreholes of small diameters are drilled into the crust, and thermometers are placed at various levels. Temperature measurements are also taken at different levels in mines.

Although heat flow patterns on the continents differ from those on the ocean floor, the average heat flow through both is similar, with a decrease in heat flow with an increase in age. The continental shields and the oldest ocean basins are colder than midocean ridges and young mountain ranges. The average global heat flow is roughly equivalent to the thermal output of a 300-watt light bulb over an area of about the size of a football field. As a comparison, the same amount of energy from the sun would cover roughly 1 square yard.

Heat flow patterns can also supplement seismic data from earthquake waves to determine the thickness of the lithosphere in various parts of the world. Seismic waves are like sound waves bouncing around inside the Earth that give a sort of X-ray picture of the Earth's interior. Once a geothermal gradient has been established for a region, it is only a matter of extrapolating the data to predict at what depth partial melting takes place in the mantle. Both seismic data and heat flow patterns show that the oceanic plates thicken as they age, from a few miles soon after formation at midocean ridges to over 50 miles in the oldest ocean basins, where the heat flow is the lowest. The reason the plates become thicker as they spread away from a midocean ridge is that material from the asthenosphere, a semimolten layer between the mantle and lithosphere, adheres to the underside of the plates and is transformed into new oceanic lithosphere. The continental plates vary in thickness from 25 miles in the young geologic provinces, where the heat flow is high, to 100 miles or more under the continental shields, where the heat flow is much lower. The shields are so thick they can actually scrape the bottom of the asthenosphere, and the drag acts like an anchor to slow down the motion of the plate.

Most of the Earth's thermal energy is generated by radioactive isotopes, mainly uranium, thorium, and potassium. The ratio of these elements to their stable daughter products also offers an accurate method for dating rocks (Figure 33). Presently, 40 percent of the heat flow at the surface is generated within the crust,

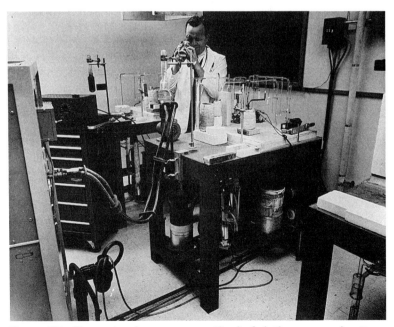

Figure 33 The potassium-argon method of dating a sample. Photo courtesy USGS

which accounts for its high geothermal gradients. The concentration of radioactive elements is greater in the crust than in the mantle, which suggests that they segregated early in the Earth's history.

When the Earth began to melt and segregate into its various layers, over twice as much heat was generated by radioactivity than is generated today. Since the formation of the Earth, half of the radioactive isotopes have decayed into stable elements. The increased heat production was probably due to a rise in heat flow on the surface. It also implies that the original lithospheric plates were thinner, smaller, and more numerous than they are today. The asthenosphere underlying the plates was probably much more active in the past. Therefore, the plates might have moved around more vigorously, causing all sorts of calamities on the Earth's surface.

TABLE 4   FREQUENTLY USED RADIOISOTOPES FOR GEOLOGIC DATING

Radioactive Parent	Half-life (years)	Daughter Product	Rocks and Minerals Commonly Dated
Uranium-238	4.5 billion	Lead-206	Zircon, uraninite, pitchblende
Uranium-235	713 million	Lead-207	Zircon, uraninite, pitchblende
Potassium-40	1.3 billion	Argon-40	Muscovite, biotite, hornblende, glauconite, sanidine, volcanic rock
Rubidium-87	47 billion	Strontium-87	Muscovite, biotite, lepidolite, microcline, glauconite, metamorphic rock
Carbon-14	5730	Nitrogen-14	All plant and animal materials

THE MANTLE

The mantle is a thick shell of red-hot rocks, separating the intensely hot metallic core from the cooler crust (Figure 34). It starts at an average depth of about 25 miles below the surface and continues to a depth of about 1800 miles. The upper mantle has a layered structure composed of a rigid lithosphere, a soft region or asthenosphere on which the rigid lithospheric plates ride, and a lower region that is in a plastic state and might be in convective flow. However, the continents are not buoyed up by the asthenosphere, which only offers passive resistance to sagging. The lower mantle is composed of rocks unchanged since the early history of the Earth. The mantle holds the majority of the Earth's materials, accounting for nearly half its radius, 83 percent of

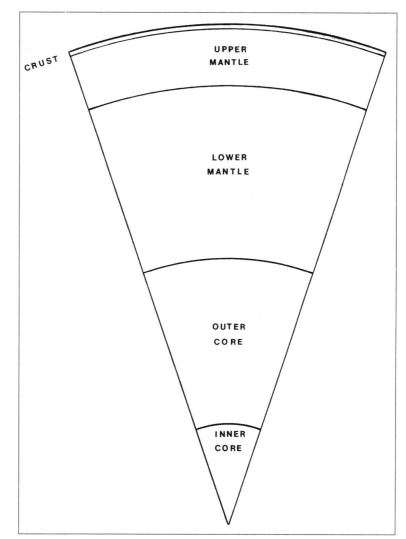

Figure 34   The layering of the Earth.

its volume, and 67 percent of its mass. The mantle is composed of iron-magnesium-rich silicate minerals, corresponding to the rock peridotite, whose name comes from the transparent green gemstone peridot.

Starting at the surface, the temperature within the Earth increases rapidly with depth. This is mostly due to the relative abundance of radioactive elements in the Earth's crust. At a depth of about 70 miles, where the minerals that make up the major constituent of the upper mantle began to melt, the temperature is about 1200 degrees Celsius at the semimolten asthenosphere. The temperature in the mantle increases gradually to about 2000 degrees at a depth of 300 miles. It then increases more rapidly to the top of the core. Between 70 and 80 percent of the mantle's heat is generated internally by radiogenic sources that produce energy by radioactive decay. The rest comes from the core.

The increase in temperature and pressure with depth results in a change in the mineral structure of the mantle rocks, called a phase transition. Phase transitions in the upper mantle occur at depths of about 45, 200, and 400 miles, corresponding to the boundaries of the lithosphere, asthenosphere, and upper mantle. The phase boundaries also correspond to the depths where earthquake wave velocities change, indicating that the upper mantle might have a layered structure. In a descending lithospheric plate, the first two phases occur at relatively shallow depths because of the plate's lower temperature. As the plate continues to descend, it changes to denser mineral forms, which help to heat the plate and speed its assimilation into the mantle.

The density of the mantle increases about 60 percent from top to bottom, and its rocks are compressible. By comparison, the crust is little more than an obscure layer of light rocks, making up only about 1 percent of the Earth's radius, and contains a thin film of ocean and atmosphere. The driving force that moves the continent around arises within the mantle, making it one powerful heat engine.

The only window scientists actually have on the mantle is through kimberlite pipes, named for the South African town of Kimberly, where they were first found. Kimberlite pipes are cores of ancient, extinct

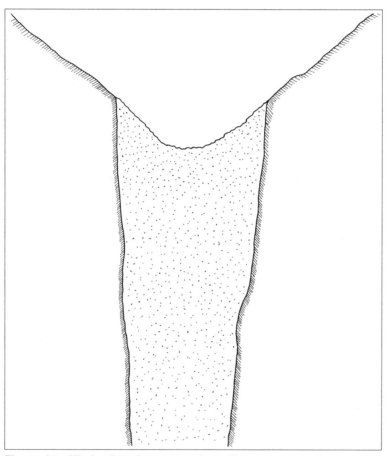

Figure 35   Kimberlite pipes are volcanolike structures that are the only window on the mantle.

volcanolike structures that extend deep into the upper mantle, as much as 150 miles below the surface, and that have been exposed by erosion. They bring diamonds to the surface from deep below by explosive eruptions. The pipes are mined extensively for these gems throughout Africa and other parts of the world. Diamonds are produced when a pure form of carbon is subjected to extreme temperature and pressure, conditions found only in the mantle. Most eco-

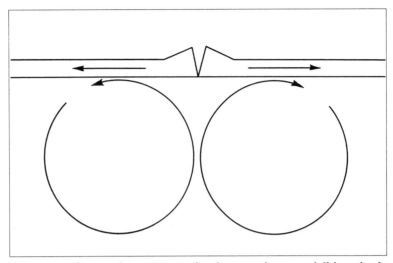

Figure 36 Convection currents in the mantle spread lithospheric plates apart.

nomic kimberlite pipes are cylindrical or slightly conical (Figure 35) and range from 1000 to 5000 feet across.

During the evolution of the Earth, beginning about 4.6 billion years ago, the mantle has had a profound influence on the conditions on the planet's surface. The mantle produced magma that rose to the surface and solidified. These basalt lava flows piled up layer by layer to form the early crust. The mantle ejected water vapor and gases, which constitute the atmosphere and ocean. The mantle also expelled carbon compounds, without which there would be no life.

The surface of the Earth was shaped by the action of the mantle moving very slowly below the crust. Without this movement, erosion would wear down mountains to the level of the sea within a space of a mere 100 million years. The surface of the Earth would then be a vast, featureless plain, unbroken by mountains and valleys. There would be no volcanoes, earthquakes, or plate tectonics. The Earth would indeed be a very uninteresting place in which to live—that is, if life could exist under these conditions.

CONVECTION CURRENTS

The mantle has now cooled to a semisolid or plastic state, except for a relatively thin layer of partially melted rock between 70 and 150 miles below the surface. This layer is called the low-velocity zone, due to an abrupt decrease in the speed of seismic waves from earthquakes traveling

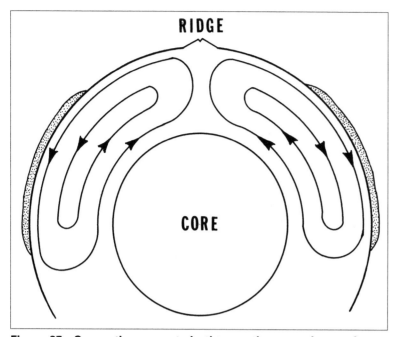

RIDGE

CORE

Figure 37 Convection currents in the mantle move the continents

through it, equivalent to the asthenosphere. The presence of water and carbon dioxide in the upper mantle acts like a catalyst that aids in the partial melting of rocks at lower pressures and makes the asthenosphere flow easily.

Heat transferred from the mantle to the asthenosphere causes convective currents to rise and travel laterally when they reach the underside of the lithosphere. Upon giving up their heat energy to the lithosphere, the currents cool and descend back into the mantle, similar to the way air currents operate in the atmosphere. If there are any cracks or areas of weakness in the lithosphere, the convective currents spread the fissures wider apart to form rift systems (Figure 36). Here the largest proportion of the Earth's interior heat is lost to the surface, as magma flows out of the rift zones to form new oceanic crust.

Convection currents transport heat by the motion of mantle material, which in turn drives the plates. The mantle convection currents are believed to originate over 400 miles below the surface. The deepest known earthquakes are detected at this level, and since almost all large earthquakes are triggered by plate motions, the energy they release must come from the forces that drive the plates. At the plate boundaries where one plate dives under another, the sinking slab meets great resistance to its motion at a depth of about 400 miles, the boundary between the upper and lower mantle.

If a slab should sink as far as the bottom of the lower mantle, however, it might provide the source material for hot spots. If all oceanic plates were to sink to this level, a volume of rock equal to that of the entire upper mantle would be thrust into the lower mantle every billion years. In order for the two mantle layers to maintain their distinct compositions, like oil floating on water, some form of return flow back to the upper mantle would be needed, and hot spot plumes seemed to fulfill this function.

Furthermore, discarded materials might be building up a slag heap at the bottom of the mantle, producing continents at the core-mantle boundary that might interfere with the geomagnetic field as well as heat flow from the core to the mantle. The shielding of heat in this manner could have a major influence on volcanism on the Earth's surface, especially its role in plate tectonics.

The mantle rocks are churning over very slowly in large-scale convection loops (Figure 37). They travel only a couple of inches a year. The convection currents might take hundreds of millions of years to complete a single loop. Some of these loops can be extremely large in the horizontal dimension, and correspond to the dimensions of the associated plate. In the case of one plate, the Pacific plate, the loop would have to reach some 6000 miles across.

In addition to these large-scale features, there may be small-scale convection cells with maximum horizontal dimensions comparable to a depth of about 400 miles, which is the thickness of the upper mantle. These smaller convection cells would act like rollers beneath a conveyer belt to propel the plates forward. Hot material rises from within the mantle and circulates horizontally near the Earth's surface, where the top 30 miles or so cools to form the rigid plates, which carry the crust around. The plates complete the mantle convection by plunging back into the Earth's interior. Thus, they are merely surface expressions of mantle convection. The convection cells might also be responsible for rising jets of lava that create chains of volcanoes, such as the Hawaiian Islands.

Another model suggests that the driving mechanism involves some 20 thermal plumes in the mantle, each several hundred miles wide. All upward movement of mantle material would then originate at the top of the core, and as the material cools near the surface it descends back to the core, where it is reheated. It is also suggested that the mantle is not viscous or fluid enough to provide convection, and the heat might simply travel directly by conduction from the hot lower level to the colder upper level.

Convection in the mantle would be expected to be strongly influenced by the Earth's rotation, similar to its influence on air and ocean currents by the Coriolis effect, which bends poleward-flowing currents to the west and equator-ward currents to the east. Yet the rotation seems to have no direct effect on the mantle. Even if there were convective flow, it might not exist in neat circular cells but instead create eddy currents, and the flow would thus become turbulent and extremely complex. Furthermore, the mantle is not only heated from below, but, like the crust, it is also heated from within by radioactive decay. This further complicates the development of convection cells and causes distortion because the interior of the cells would no longer be passive but provide a significant portion of the heat as well. There is laboratory evidence that under certain conditions mantle convection might extend downward to the core.

Since slabs of sinking material meet great resistance at a depth of about 400 miles, convection would have to be limited to the region above this depth. Yet there must be large-scale convections to account for the geophysical observations, especially the motion of the surface plates. It might be that large-scale motions are superimposed on small-scale convective features similar to the circulation of the atmosphere and oceans. To test this hypothesis, there must be geophysical evidence that small-scale flow does indeed exist.

Because the plates are so rigid, they tend to mask most of the effects that might be associated with small-scale convection. Two important effects that can be observed, however, are the variation in gravity and greater-than-expected heat flows on the ocean floor. Variances in gravity are usually due to differences in density of the mantle and associated deformations on the surface, such as variations in the depth of the ocean. This results from the cooling and shrinking of a plate as it moves away from a midocean ridge, where the growth of the plate begins with upwelling of hot material from the mantle. Usually a large magma chamber lies under a ridge where the lithosphere is created at a fast rate, and a narrow magma chamber lies under a slow-spreading center.

Also, because all plates cool in much the same manner, the depth of the ocean should be a function of age and can be readily calculated. Regional departures from the expected depth correspond quite closely to the gravity variations. There are good reasons to believe the variations in gravity and depth are associated with convection currents operating at the base of the plate. Chains of volcanoes, such as the Hawaiian Islands, do not move away from their sources of volcanism as fast as the plates on which they ride. Therefore, the sources must be in motion, indicating that convection currents are distorting the flow of upwelling mantle plumes that contain molten mantle material originating from sources as deep as just above the core.

SUPERCONTINENTS

Throughout the Earth's history, the continents have undergone the process of collision and rifting on a grand scale. Separate blocks of continental crust collided and merged into larger continents. Later, the continents were torn apart by deep rifts that eventually became new oceans. The convective motions in the mantle that drive continents around the face of the Earth are powered by heat generated by the decay of radioactive elements. The Earth's internally generated radiogenic heat is steadily declining, however. Therefore, it alone cannot be responsible for the continuing, alternating cycle of assembly and breakup of continents.

The key phenomenon in this process is not the production of heat but the conduction and loss of heat through the crust. Continental crust is only half as efficient as oceanic crust at conducting heat and acts like an insulating thermal cap. A supercontinent covering a portion of the Earth's surface, where heat from the mantle can accumulate under the crust, causes it to dome upward, creating a superswell. New continents eventually rift apart, slide off the superswell, and move toward colder sinking regions in the mantle. Individual continents then become trapped over separate, cool downflows of mantle rock.

After the continents have been widely dispersed, heat is more easily conducted through the newly formed ocean basins. When a certain amount of heat has escaped, the continents halt their progress and start to return toward each other. When the present continents have reached their maximum dispersal, the crust of the Atlantic will age and become dense enough to sink under the surrounding landmasses, thus beginning the process of closing the Atlantic Basin. As the convection patterns shift, all continents will rejoin into a supercontinent over a large downflow, and a new cycle begins.

The rifting of a continent brought about by hotspot volcanism at rift valleys takes place on average about once every 30 million years. The rising of mantle material in giant plumes underplates the crust with basaltic magma, which heats and weakens it, causing huge blocks to downdrop and form a series of grabens, or down-faulted blocks. Convection currents in the mantle pull the thinning crust apart, forming a deep rift valley similar to the East African Rift system (Figure 37a). As the crust gets thinner, volcanic eruptions become more prevalent, further weakening the crust.

Figure 37a   The East African Rift system, where the continent is being pulled apart by plate tectonics.

The rift then fills with seawater after a breach has opened up to the ocean. Eventually, the continental rift is replaced by a midocean ridge spreading system, and the creation of new ocean floor between the two pieces of crust pushes them further apart. In this manner, over 2 million cubic miles of molten rock are released every time a supercontinent rifts apart near a hot plume. After the breakup, the continents travel in spurts rather than drift at a constant speed.

The regions bordering the Pacific Ocean apparently have not undergone continental collisions. In effect, the Pacific is a remnant of an ancient sea that has narrowed and widened in response to the drifting of the continents. The Pacific plate, which now covers one quarter of the Earth's surface, was hardly larger than the United States after the breakup of Pangaea. The rest of the ocean floor consisted of other, unknown plates that have since disappeared as the Pacific plate grew. Oceans have repeatedly opened and closed in the vicinity of the North Atlantic, while a single ocean has existed continuously in the vicinity of the Pacific. The Pacific plate is being subducted under all the continents that surround it, whereas the oceanic crust in the Atlantic butts against the surrounding continents.

As the material on the ocean floor ages, it cools, becomes denser, and subsides, increasing the depth of the ocean. Eventually about 200 million years after the first rift formed, the oldest part of the new ocean floor adjacent to the continent becomes so dense that it sinks under the continental crust and is subducted into the mantle. The process of subduction closes the ocean, bringing the continent back together.

When continents rejoin, the compressive forces of the collision form mountain belts. The ages of mountain ranges produced by continental collisions are remarkable in their regularity. Intense mountain building occurred about 2.6 billion years ago, 2.1 billion years ago, 1.8 to 1.6 billion years ago, 1.1 billion years ago, 650 million years ago, and 250 million years ago. The timing of these episodes shows an apparent periodicity of between 400 million and 500 million years.

About 100 million years after each of these events, there appears to have been a period of rifting, occurring about 2.5 billion years ago, 2 billion years ago, 1.7 to 1.5 billion years ago, 1 billion years ago, 600 million years ago, and 150 million years ago. Continents separate some 40 million years after rifting and take about 160 million years to reach their greatest dispersal and for subduction to begin in the new oceans. After the continents begin to move back together, another 160 million years might elapse before they re-form a new supercontinent. The supercontinent might survive for about 80 million years before it rifts apart, changing once again the face of the Earth.

4

CRUSTAL PLATES

Not even the largest and most prominent features on the Earth's surface can be regarded as permanent and immovable. The Earth's crust comprises rigid lithospheric plates, and because they are constantly in motion, continents and oceans are continuously being reshaped and rearranged. It is becoming more apparent that plate tectonics and continental drift have played a prominent role in climate and life. Changes in continental configurations brought on by movable crustal plates have profoundly affected global temperatures, ocean currents, biologic productivity, and many other factors of fundamental importance to the living Earth.

THE EARTH'S CRUST

The bulk of the Earth's crust is composed of oxygen, silica, and aluminum, which form the granitic rocks that constitute most of the continents. The crust and the upper brittle mantle constitute the lithosphere, which averages about 60 miles thick. The lithosphere rides freely on the semimolten outer layer of the mantle called the asthenosphere, which ranges in depth from about 70 to 150 miles. This gives the outer layer of the Earth a structure somewhat like a jelly sandwich, which is important for the operation of plate tectonics. Otherwise, the crust would be just jumbled up slabs of rock, and the Earth would be an alien place indeed.

TABLE 5   COMPOSITION OF THE EARTH'S CRUST

Crust Type	Shell	Average Thickness (in miles)	Percent Composition of Oxides						
			Silica	Alum	Iron	Magn	Calc	Sodi	Potas
Continental	Sedimentary	2.1	50	13	6	3	12	2	2
	Granitic	12.5	64	15	5	2	4	3	3
	Basaltic	<u>12.5</u>	58	16	8	4	6	3	3
Total		27.1							
Sub-continental	Sedimentary	1.8							
	Granitic	5.6			Same as above				
	Basaltic	<u>7.3</u>							
Total		14.7							
Oceanic	Sedimentary	0.3	41	11	6	3	17	1	2
	Volcanic	0.7	46	14	7	5	14	2	1
	Sedimentary Basaltic	<u>3.5</u>	50	17	8	7	12	3	<1
Total		4.5							
Average		15.4	52	14	7	4	11	2	2

The Earth's crust is relatively thin compared with that of the Moon and the other terrestrial planets, Mercury, Venus, and Mars, which have thick, buoyant, nonsubducting crusts because they are either too cold or too hot, and thus have been tectonically inactive for over 2 billion years. A thick buoyant crust on Earth would have remelted because of the high concentration of radioactive elements and great pressures induced by the weight of the overlying rocks. Such a thick crust would have acted like an insulating blanket to hold in the constantly generated heat from the Earth's interior, raising the internal temperature high enough to melt surface rocks.

A thick crust would also have been highly unstable, creating a massive overturn that would melt the entire crust. Since there is no vestige of the first 700 million years of Earth history, this might attest to such an occurrence.

A thick, buoyant crust could not be easily broken up and subducted into the mantle, which is important for the operation of global tectonics. The lithospheric plates would simply float on the surface like pack ice in the Arctic. This would indeed make the Earth an uninteresting place, for there would be no majestic mountains, no deep blue seas, no volcanoes and earthquakes, and no life—at least not as we know it.

The continental crust averages 25 to 30 miles thick and is up to 45 miles thick in the mountainous regions. The oceanic crust, however, is considerably thinner and in most places is only 3 to 5 miles thick. Like an iceberg, only the tip of the crust shows, while the rest is out of sight deep below the surface. The continental crust is 20 times older than the oceanic crust, which is found nowhere older than 170 million years. This is because the older oceanic crust has been consumed by the mantle at subduction zones spread around the world. (See Chapter 6.) Because of plate tectonics, perhaps as many as 20 oceans have come and gone during the last 2 billion years.

Eight major and about a half dozen minor lithospheric plates (Figure 38) act as rafts that carry the crust around on a sea of molten rock. The plates diverge at midocean ridges and converge at subduction zones, which are expressed on the ocean floor as deep-sea trenches, where the plates are subducted into the mantle and remelted. The plates and oceanic crust are continuously recycled through the mantle, but the continental crust because of its greater buoyancy remains for the most part on the surface.

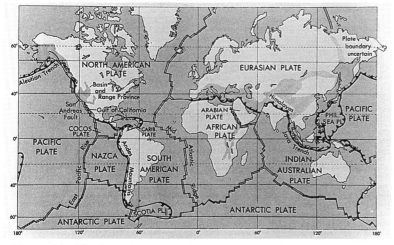

Figure 38   The Earth's crust is fashioned out of several lithospheric plates, identified here, whose motions are responsible for all geologic activity. Courtesy USGS

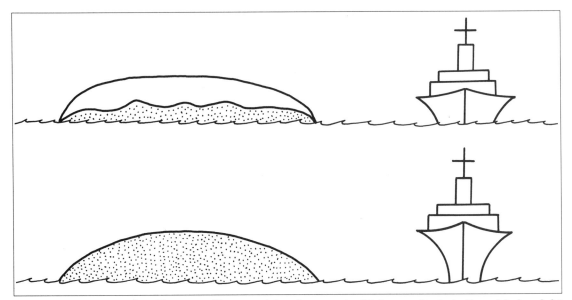

Figure 39   The principle of isostasy, illustrated. Land covered with ice readjusts to the added weight like a loaded freighter. When the ice melts, the land is buoyed upward as the weight decreases.

An interesting feature about the Earth's crust that geologists found quite by accident was that Scandinavia and parts of Canada are slowly rising nearly half an inch a year. Over the centuries, mooring rings on harbor walls in Baltic seaports have risen so far above sea level they could no longer be used to tie up ships. During the last Ice Age, which ended about 10,000 years ago, the northern landmasses were covered with ice sheets up to 2 miles thick. Under the weight of the ice, North America and Scandinavia began to sink like an overloaded ship. When the ice melted, the crust became lighter and began to rise (Figure 39).

In Scandinavia, marine fossil beds have risen more than 1000 feet above sea level since the last ice age. The weight of the ice sheets depressed the landmass when the marine deposits were being laid down. When the ice sheets melted, the removal of the weight raised the landmass due to its greater buoyancy. This effect is responsible for maintaining equilibrium in the Earth's crust. Therefore, the lighter continents acted as though they floated on a sea of heavier rocks.

It is due to plate tectonics that life was able to flourish on Earth. It is even possible that there would not be active plate tectonics if Earth did not possess life. Lime-secreting organisms in the ocean remove carbon dioxide, an important greenhouse gas, from the atmosphere and store it in the bottom sediments. This keeps the Earth's surface temperature within the range needed for plate tectonics to operate, which in turn maintains living conditions on Earth.

TABLE 6   CLASSIFICATION OF THE EARTH'S CRUST

Environment	Crust Type	Tectonic Character	Crustal Thickness (miles)	Geologic Features
Continental crust overlying stable mantle	Shield	Very stable	22	Little or no sediment, exposed Precambrian rocks
	Mid-continent	Stable	24	
	Basin range	Very unstable	20	Recent normal faulting, volcanism, and intrusion; high mean elevation
Continental crust overlying unstable mantle	Alpine	Very unstable	34	Rapid recent uplift, relatively recent intrusion; high mean elevation
	Island arc	Very unstable	20	High volcanism, intense folding and faulting
Oceanic crust overlying stable mantle	Ocean basin	Very stable	7	Very thin sediments overlying basalts, no thick Palaeozoic sediments
Oceanic crust overlying unstable mantle	Ocean ridge	Unstable	6	Active basaltic volcanism, little or no sediment

The lithosphere, which includes the crust and its underlying plate, is generally between 50 and 100 miles thick under the continents, whereas beneath the ocean floor it ranges from about 5 miles thick near spreading centers to about 60 miles thick at plate margins. Cracking open the continental lithosphere would appear to be a formidable task, considering its great thickness. The best evidence for the rifting of continents can be found in the East African Rift system, which has not yet fully ruptured. When it

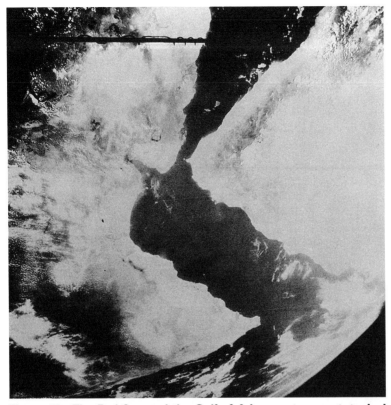

Figure 40 The Red Sea and the Gulf of Aden are two prototypical oceans created by seafloor spreading. Photo from Earthquake Information Bulletin 441, courtesy USGS

does, however, the present continental rift will be replaced by an oceanic rift as the area floods with seawater. This type of rifting is presently taking place in the Red Sea, where Africa and Saudi Arabia are slowly drifting away from each other (Figure 40).

The transition from a continental rift to an oceanic rift is accomplished by block faulting, in which blocks of continental crust called grabens, drop down along extensional faults, where the crust is being pulled apart, resulting in a deep rift valley and a thinning of the crust. Upwelling of molten rock from the mantle further weakens the crust, and convection currents pull the crust apart.

As rocks heat up in the asthenosphere, they become plastic, slowly rise by convection, and after millions of years reach the topmost layer of the mantle, or lithosphere. When the rising rocks reach the underside of the lithosphere, they spread out laterally, cool, and descend back into the deep interior of the Earth. The constant pressure against the bottom of the lithosphere creates fractures that weaken it. As the convection currents flow out on either side of the fracture, they carry the two separated parts of the lithosphere along with them and the rift continues to widen. Eventually, the continental rift is replaced by an oceanic rift when the area is flooded with seawater.

With reduced pressure, the rocks melt and rise up through the fracture. The molten magma passes through the 60 miles or so of lithosphere until it reaches the oceanic crust, where it forms magma chambers that further press laterally against the oceanic crust and widen it. The magma chambers also provide molten lava that pours out from the trough between the two ridge crests, adding layer upon layer to both sides of the spreading ridge.

This generates about 3 cubic miles of new oceanic crust every year. The pressure of the upwelling magma forces the ridge farther apart, pushing the ocean floor and the lithosphere upon which it rides away from the mid-ocean ridge.

TERRANES

The continents are composed of cratons, the odds and ends of ancient continental crust. The cratons are patchwork mixtures, consisting of crustal pieces known as terranes, which are assembled into geologic collages. The terranes are usually bounded by faults and are distinct from their geologic surroundings. The composition of terranes generally resembles that of an oceanic island or plateau. Some, however, are composed of a consolidated conglomerate of pebbles, sand, and silt that accumulated in an ocean basin between colliding crustal fragments.

The entire state of Alaska is an agglomeration of terranes (Figure 41). A large portion of the Alaskan Panhandle, known as the Alexander terrane was once part of eastern Australia some 500 million years ago. It broke off from Australia, traversed the Pacific Ocean, stopped briefly at the coast of Peru, sliced past California, swiping some of the Mother Lode, and bumped into North America around 100 million years ago.

The actual distances that terranes travel vary considerably. Basaltic seamounts that accreted to the margin of Oregon moved from nearby offshore, while similar rock formations around San Francisco, California, (Figure 42) came from as far away as 2500 miles across the Pacific Ocean. At their usual rate of travel, terranes could make a complete circuit of the globe in only about 500 million years.

Terranes come in a variety of shapes and sizes

Figure 41 The Brooks Range near the head of the Itkillik River east of Anaktuvuk Pass, showing steeply tilted Paleozoic rocks. Photo 884 by J. C. Reed, courtesy USGS

from small slivers to subcontinents such as India, which is a single great terrane. Most terranes are elongated bodies that tend to become deformed when they collide and accrete to a continent. The assemblage of terranes in China is being stretched and displaced in an east-west direction due to the continuing squeeze India is exerting on southern Asia. The terranes range in age from less than 200 million years old to well over a billion years old.

HOT SPOTS

Over 100 small regions of isolated volcanic activity known as hot spots are found in various parts of the world (Figure 43). Unlike most other active volcanoes, hot-spot volcanoes rarely exist at plate boundaries but instead lie deep in the interior of a plate. They are notable if only for their very isolation, far removed from normal centers of volcanic and earthquake activity, and might be the only distinctive feature on an otherwise monotonous landscape.

Almost all hot-spot volcanism occurs in regions of broad crustal uplift or swelling. Lavas of hot-spot volcanoes differ markedly from those of rift systems and subduction zones. The lavas are composed of basalts containing larger amounts of alkali minerals such as sodium and potassium, indicating that their source material is not associated with plate margins.

Hot spots appear to result from plumes of hot material rising from deep within the mantle, possibly just above the core. The distinctive composition of hot-spot lavas seems to indicate a source outside the general circulation pattern of the man-

Figure 42   The San Francisco Bay area from _Landsat_ Earth resource satellite. Photo courtesy NASA

tle. Plumes might also arise from stagnant regions in the center of convection cells or from below the region in the mantle that is stirred by convection currents. As plumes of mantle material flow upward into the atmosphere, the portion rich in volatiles rises toward the surface to feed hot-spot volcanoes. The plumes come in a range of sizes that might indicate the depth of their source material. The plumes are not necessarily continuous flows of mantle material, but might consist of molten rock, rising in giant blobs or diapirs.

Most hot spots move very slowly, slower than the oceanic or continental plates above them. When a continental plate hovers over a hot spot, the molten magma welling up from deep below creates a broad domelike structure in the crust that averages about 125 miles across and accounts for about 10 percent of the Earth's total surface area. As the dome grows, it develops deep fissures through which magma can find its way to the surface.

Often the passage of a plate over a hot spot re-

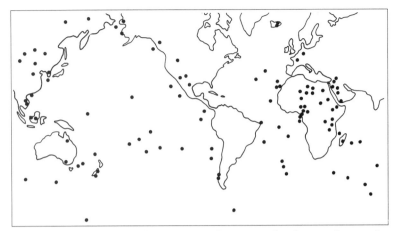

Figure 43 The world's hot spots, where mantle plumes rise to the surface.

Figure 44 Photograph of the Hawaiian Island chain, taken looking south from the space shuttle. The main island Hawaii is in the upper portion of the photograph. Photo courtesy NASA

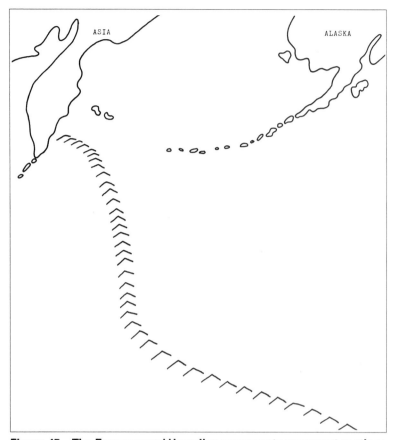

Figure 45   The Emperor and Hawaiian seamounts represent motions in the Pacific plate over a volcanic hot spot deep within the mantle.

sults in a trail of volcanic features, whose linear trend reveals the direction of plate motion. This produces volcanic structures aligned in a direction that is oblique to the adjacent midocean ridge system rather than parallel to it like rift volcanoes. The hot-spot track might be a continuous volcanic ridge or a chain of volcanic islands and seamounts that rise high above the surrounding seafloor. The hot-spot track might also weaken the crust, cutting through the lithosphere like a geologic blowtorch.

The most prominent and most easily recognizable hot spot created the Hawaiian Islands (Figure 44). Apparently all the islands in the Hawaiian chain were produced by a single source of magma, over which the Pacific plate has passed, proceeding in a northwesterly direction. The volcanic islands popped out on the ocean floor in conveyer belt fashion, with the oldest trailing off to the northwest, farthest from the hot spot. There are also similar chains of volcanic islands in the Pacific that trend in the same direction as the Hawaiian Islands. This effect indicates that the Pacific plate is moving off in the direction defined by the volcanic chains.

The trail of volcanoes left by the hot spots changes abruptly to the north, where it follows the Emperor Seamounts, the isolated undersea volcanoes strung out in a chain across the interior of the Pacific plate (Figure 45). This deflection occurred about 40 million years ago, the same time India rammed into Asia, which might have shifted the Asian plate. Therefore, hot spots could be a reliable means for determining the direction of plate motion, provided they remain reasonably stationary.

More than half of the hot spots exist on the continents, with the greatest concentration, about 25 in all, in Africa (Figure 46), which has remained essentially stationary for millions of years. Hot spots might have been responsible for the unusual topography of the African continent, which is characterized by numerous basins, swells, and uplifted highlands. The effect might also indicate that the African plate has come to rest over a population of hot spots. Further evidence that Africa is stationary is that hot-spot lavas of several different ages are superimposed on one another. If the continent were drifting, the hot-spot lavas would spread laterally in a chronological sequence.

There also appears to be a direct relationship between the number of hot spots and the rate of drift of a continent. Besides Africa, hot spots are numerous in Antarctica and Eurasia, so it would seem that these regions are moving at a very slow place as well. In contrast, on rapidly moving continental plates, such as North and South America, hot-spot volcanism is rare.

Yellowstone National Park is more than 1000 miles from the nearest plate boundary, yet it is one of several midplate centers of hot-spot activity. Beneath the park lies a hot spot that is responsible for the continuous thermal activity giving rise to a number of geysers such as Old Faithful (Figure 47). The geysers are produced when water seeps into the ground, is heated near a magma chamber, and rises explosively through fissures in the torn crust.

The hot spot was not always under Yellowstone, however, and its positions relative to the North American plate can be traced through volcanic rocks on the Snake River Plain for 400

Figure 46 The Brandberg structure in Namibia near Cape Cross in Southwest Africa is a zone of weakness in the Earth's crust created by upwelling magma. Photo courtesy NASA

Figure 47   Eruption of Old Faithful Geyser, Yellowstone National Park, Wyoming. Photo courtesy USGS

miles in southern Idaho. Over the past 15 million years, the North American plate slid southwestward across the hot spot, placing it under its temporary home at Yellowstone. During the past 2 million years, there have been at least three episodes of intense volcanic activity in the region, and an explosive eruption hundreds of times greater than Mount St. Helens is well overdue. Eventually, as the plate continues in its westerly direction, the relative motion of the hot spot will bring it across Wyoming and Montana.

Sometimes a hot spot fades away entirely, and a new one forms in its place. The typical life span of a plume is on the order of about 100 million years. The position of a hot spot can change slightly as it sways in the convective currents of the mantle. As a result, the tracks on the surface might not always be as linear as those of the Hawaiian chain. But compared with the motion of the plates, the mantle plumes are relatively stationary. Because the motion of the hot spots is insignificant, they provide a reference point for determining the direction and rate of plate travel. If the upwelling plumes should cease erupting, the plates would grind to a complete halt due to the loss of internal heat sources.

PLATE MOTIONS

The plates are composed of the lithosphere and the overlying continental or oceanic crusts. The plate boundaries are midocean ridges, subduction zones, and transform faults. The plates ride on the asthenosphere and carry the continents along with them like ships frozen in floating ice sheets. The breakup of a plate results in the formation of a new continent such as the breakup of Laurasia and Gondwana at the Mid-Atlantic Ridge to create North and South America and Eurasia and Africa. On the other hand, if two plates collide, the thrust of one plate under the other uplifts the crust to form mountain ranges or long chains of volcanic islands.

Thousands of feet of sediments are deposited along the seaward margin of a continental plate in deep-ocean trenches, and the increased weight presses downward on the oceanic crust. As the continental and oceanic plates merge, the heavier oceanic plate is subducted or overridden by the lighter continental plate, forcing it further downward. The sedimentary layers of both plates are squeezed, causing a swelling at the leading edge of the continental crust, forming mountain belts. The sediments are faulted at or near the surface, where the rocks are brittle, and folded at depths where the rocks are more plastic.

As the oceanic crust is descending, the topmost layers are scraped off and are plastered against the swollen edge of the continental crust, forming an accretionary wedge. In the deepest part of the continental crust, where temperatures and pressures are very high, rocks are partially melted and metamorphosed. As the descending plate dives farther under the continent, it reaches depths where the temperatures are extremely high. Part of it melts forming a silica-rich magma that rises because it is lighter than the surrounding rock material. The magma intrudes the overlying metamorphic and sedimentary layers to form large bodies of granite or erupts on the surface from a volcano.

Continents are neither created nor destroyed by the process of plate tectonics, only the plates on which they ride are absorbed. Divergence of lithospheric plates creates new oceanic crust, whereas convergence destroys oceanic crust in well-developed subduction zones. This condition is prevalent in the western Pacific, where deep subduction zones are responsible for island arcs. Volcanoes of the island arcs are highly spectacular because their lava is silica-rich, contrasting strongly with the basalt of other volcanoes and midocean ridges. The volcanoes are mostly explosive and build up steep-sided cinder cones. Island arcs are also associated with belts of deep-seated earthquakes 200 to 400 miles below the surface.

Rifts open not only in ocean basins but also under continents. This is happening in eastern Africa, creating a great rift valley that will eventually widen and flood with seawater to form a new subcontinent. Old extinct rift systems, where the spreading activity has stopped, or failed rifts, where a full-fledge spreading center did not develop, are overrun by continents. For example, the western edge of North America has overrun the northern part of the now extinct Pacific rift system. The North American continental mass has run into the northern extension of the active Pacific rift system, called the East Pacific Rise, creating the San Andreas Fault in California.

Measuring the rate of plate motions requires an extreme accuracy over a distance of thousands of miles. Standard geodetic survey methods cannot provide this accuracy, but measurements using satellites can. In Satellite Laser Ranging, distances are measured by comparing how long it takes for laser pulses to leave the surface, bounce off a satellite, and return to ground stations. In the satellite-based Global Positioning System, plate positions can be measured with a precision of about 1 inch over a distance of 300 miles.

In very-long-baseline interferometry (VLBI), radio signals from distant quasars (rapidly spinning collapsed stars) are monitored at different stations on Earth. The difference in arrival times for these signals determine the distance between the receiving stations. The VLBI method is the more exact of these techniques with an accuracy approaching several parts per billion, comparable to measuring the length of 100 football fields to within the width of a human hair. These measurements of plate motion are in good agreement with geologic methods, which are based on the spacing of magnetic stripes on the ocean floor produced by seafloor spreading.

EARTHQUAKES

Earthquakes are by far the strongest natural forces on Earth; in a matter of seconds, a large quake can level an entire city. Every year, a dozen or so major earthquakes strike somewhere in the world. Most of these are in areas along the rim of the Pacific plate. The Pacific Basin area is also prone to destructive seismic sea waves or tsunamis from undersea earthquakes (Figure 48).

The mechanism for creating earthquakes was poorly understood until after the big quake that struck San Francisco in 1906 (Figure 49). The American geologist Harry Reid discovered that for hundreds of miles along the San Andreas Fault, fences and roads crossing the fault had been displaced by as much as 21 feet. This led him to propose the modern elastic-rebound theory of earthquake faulting.

The San Andreas Fault (Figure 50) is a fracture zone 650 miles long and 20 miles deep that runs

Figure 48 **Tsunamis washed many vessels into the heart of Kodiak from the March 27, 1964, Alaskan earthquake.** Photo courtesy USGS

northward from the Mexican border through southern California and plunges into the Pacific Ocean just south of the Oregon border. The fault represents the margin between the Pacific plate and the North American plate, which are moving relative to each other in a right lateral direction at a rate of nearly 2 inches per year. The fault absorbs most of this motion, and the rest is dissipated by the spreading of the Basin and Range Province and by the deformation of California's southern coastal ranges. In the 50 years before the great San Francisco earthquake, land surveys showed displacements as much as 10 feet along the San Andreas Fault. Tectonic forces slowly deformed the crustal rocks on both sides of the fault, causing large displacements. All this time, the rocks were bending and storing up elastic energy, like stretching a rubber band.

Eventually, however, the forces holding the

Figure 49   Secondary cracks in pavement from the 1906 San Francisco earthquake.   Photo 2871 by G. K. Gilbert, courtesy USGS

rocks together are overcome. Slippage occurs at the weakest point, and, like a rubber band, the rocks snap back. The point of initial rupture is called the hypocenter, and, if near the surface, it can cause large displacements in the crust. These displacements can exert strain further along the fault, where additional slippage can occur until most of the built-up strain is released. The slippage allows the deformed rock to elastically rebound to

Figure 50 **Southern view of the San Andreas Fault in the Carrizo Plains, California.** Photo 217 by R. E. Wallace

its original shape, which releases heat generated by friction and produces vibrations called seismic waves. The seismic waves radiate outward from the hypocenter in all directions, like the ripples produced when a rock is thrown into a quiet pond (Figure 51a).

The rocks do not always rebound immediately, however. They might take days or even years, resulting in an aseismic slip, and the seismic energy thus produced is then quite small. Why seismic energy is released violently in some cases and not in others is still not fully understood. Moreover, some types of shallow earthquakes might be triggered by outside events such as large meteorite impacts. It is even suspected that when the gravitational attractions of the sun and the moon pull together on the Earth, they can cause some earthquake fault systems to rupture. The magnitude of an earthquake is recorded by a seismograph and measured on the Richter scale, which was devised by American seismologist Charles F. Richter. The magnitude scale is logarithmic, and an increase of 1 magnitude signifies a 10-fold increase in ground motion and about 30 times the energy. A 3 on the scale is barely perceptible and 9 or more is catastrophic.

The San Andreas Fault is perhaps the best-studied fault system in the world. Various remote-sensing techniques are employed for making earthquake predictions. Laser ranging devices measure the amount of crustal strain along the fault with an accuracy of 0.5 inch over a distance of about 20 miles. There are several precursory signals that faults give out that might also aid in earthquake prediction. These include changes in the tilt of the ground, magnetic anomalies, increased radon gas content in nearby water wells, and swarms of microearthquakes.

Faults also produce a phenomenon known as earthquake lights before and during rupture (Figure 51b). Apparently, the strain on the rocks in the vicinity of the fault causes them to emit energy, producing a faint atmospheric glow at night. Seismic activity is associated with a variety of other electrical effects that might aid in predicting earthquakes. A system of radio wave monitors distributed along the San Andreas Fault have recorded changes in atmospheric radio waves prior to several earthquakes, includ-

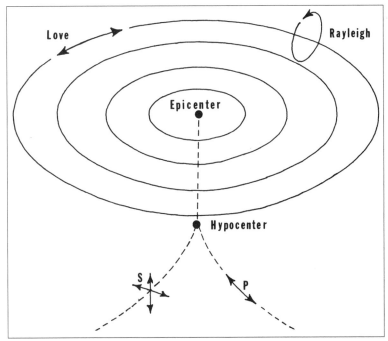

Figure 51a   Types of earthquake waves, illustrated.

Figure 51b   Earthquake lights during the Matsushiro earthquake swarm in Japan, which lasted from 1965 to 1967. Photo by T. Kuribayashi, courtesy USGS

ing the 1989 Loma Prieta earthquake. The increased electrical conductivity of rocks under stress near the fault apparently causes radio waves to be absorbed by the ground one to six days before an earthquake. Moreover, investigators have observed short pulses of increased radio interference caused by the release of electromagnetic energy by cracking rocks.

Over the past few years, geodesists (scientists who measure the Earth) have significantly improved the precision with which they can determine positions on the Earth's surface, using the satellite-based Global Positioning System (Figure 52). With this system, investigators can monitor the strain accumulation on the San Andreas Fault more efficiently than by standard geodetic methods. Positions are determined with an accuracy of about 1 part in 10 million over a distance of up to 300 miles. The orbit of the satellite is calculated with great precision by comparing satellite signals received at two stations on the ground, and this in turn provides data on their relative positions.

In addition to monitoring the San Andreas Fault, the satellites can measure the rate of seafloor spreading in Iceland, subsidence of the crust due to the removal of groundwater, and the bulging crust above the magma chamber under Long Valley, California. The Global Positioning System might also be able to detect slight bulges in the crust before a volcanic eruption.

Figure 52 An artist's conception of one of the 24 satellites of the Global Positioning System. Courtesy U.S. Air Force

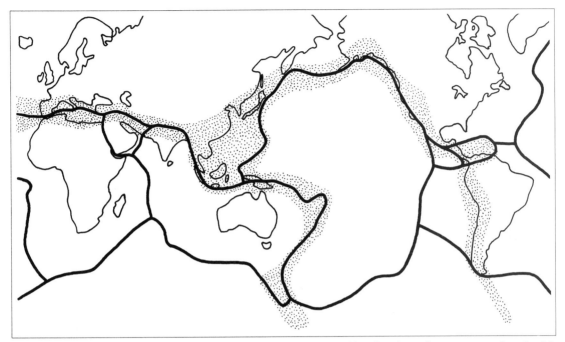

Figure 53   Most earthquakes occur in broad zones, indicated by the dotted areas, associated with plate boundaries.

FAULT ZONES

Roughly 95 percent of the seismic energy released by earthquakes is concentrated in narrow zones that wind around the globe and are associated with plate boundaries (Figure 53). A continuous belt extends for thousands of miles through the world's oceans and coincides with the midocean rift systems. Earthquakes are also associated with terrestrial rift zones such as the 3600-mile East African Rift.

The greatest amount of energy is released along a path located near the outer edge of the Pacific Ocean, known as the circum-Pacific belt. This includes the San Andreas Fault in southern California, which was responsible for numerous powerful earthquakes, including the October 1989 Loma Prieta earthquake of 7.1 magnitude, which damaged the San Francisco area (Figure 54).

Another belt runs through the folded mountainous regions that flank the Mediterranean Sea. It continues through Iran and past the Himalayan Mountains into China. A massive earthquake in the Caspian Sea just north of Iran killed 100,000 people and left half a million homeless in June 1990.

In the eastern Himalaya range lies perhaps the most seismically active region of the world. An immense seismic belt some 2500 miles long stretches across Tibet and much of China, where the 1976 Tangshan earthquake killed nearly half a million people. For centuries, this area has been shaken by catastrophic earthquakes responsible for the deaths of millions. In this century alone, more than a dozen earthquakes of 8.0 magnitude or greater have been recorded in this region.

West of this belt, in the Hindu Kush range of north Afghanistan and the nearby Russian Republic of Tadzhikistan, is the seat of many earthquakes. Three earthquakes in this century have had magnitudes of 8.0 or over. The great 1988 Armenian earthquake killed 25,000 people and left a million more homeless. This is a notoriously active seismic belt, with some 2000 minor earthquakes registered annually.

From there, the Persian arc spreads in a wide sweep through the Pamir and Caucasus Mountains and on to Turkey. The eastern end of the Mediterranean is a jumbled region of colliding plates, providing highly unstable ground. The whole of the Near East is inherently unstable, attesting to the many earthquakes reported in Biblical times. The remaining regions surrounding the Mediterranean have been devastated by earthquakes throughout history.

Figure 54 Collapsed bridge on Interstate 880 in San Francisco, a result of the October 17, 1989, Loma Prieta Earthquake, Alameda County, California. Photo by G. Plafker, courtesy USGS

The circum-Pacific belt coincides with the ring of fire because the same tectonic forces that produce earthquakes are also responsible for volcanic activity (See chapter 8). The area of greatest seismicity is on the plate boundaries associated with deep trenches and volcanic island arcs, where an ocean plate is thrust under a continental plate. Japan, which is in the process of being plastered against Asia, is a constant reminder of powerful earthquakes associated with subduction zones. The great 1923 Tokyo earthquake of 8.3 magnitude took about 140,000 lives.

The Andes Mountain regions of Central and South America, especially in Chile and Peru, are known for some of the largest and most destructive earthquakes in historic times. In this century alone, nearly two dozen earthquakes of 7.5 magnitude or greater have taken place in Central and South America, including the largest ever recorded, the great 1960 Chilean earthquake of 9.5 magnitude.

TABLE 7   SUMMARY OF EARTHQUAKE PARAMETERS

Magnitude	Surface Wave Height (feet)	Length of Fault Affected (miles)	Diameter of Area Quake Is Felt (miles)	Number of Quakes per Year
9	Largest earthquakes ever recorded—between 8 and 9			
8	300	500	750	1.5
7	30	25	500	15
6	3	5	280	150
5	0.3	1.9	190	1,500
4	0.03	0.8	100	15,000
3	0.003	0.3	20	150,000

The whole western seaboard of South America is affected by an immense subduction zone just off the coast. The lithospheric plate on which the South American continent rides is forcing the Nazca plate to buckle under, causing great tensions to build up deep within the crust. The plate is being consumed by the Peru-Chile trench at a rate of about 50 miles every million years. While some rocks are being forced deep down, others are pushed upward toward the surface, raising the Andean mountain chain. The resulting forces are building great stresses into the entire region. When the stresses become large enough, earthquakes crack open the crust, forming tall scarps that slice across the countryside.

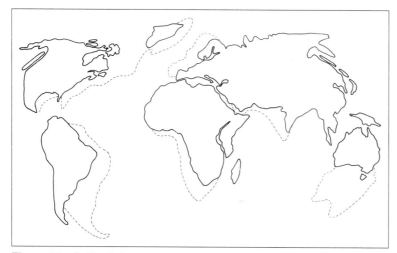

Figure 55   Drift of the continents 50 million years from now. Dashed lines indicate present positions of the continents.

Even in the so-called stable zones, earthquakes can and do occur, although not nearly as frequently as in the earthquake-prone areas. The stable zones are generally associated with continental shields, which are composed of ancient granitic and metamorphic rocks lying in the interior of the continents. When earthquakes do occur in these regions, they might be due to the weakening of the crust by compressive forces that originate at plate edges. The underlying crust might also have been weakened by previous tectonic activity, resulting in the sudden release of pent up stresses.

As time goes by, the Atlantic Ocean will continue to widen at the expense of the Pacific (Figure 55). The California coast will break off at the San Andreas Fault and proceed northward, where the plate on which it is riding will plunge down the Aleutian Trench. North America will continue to slide across the Pacific and eventually reverse direction and head back toward Eurasia. Meanwhile, South America will separate from North America and drift into the South Pacific. The African and Eurasian plates will continue to press against each other, and the Mediterranean Sea, caught in the middle, will be squeezed dry. A new subcontinent will tear off of eastern Africa and probably drift into India. Australia will drift northward, possibly colliding with Southeast Asia.

About 200 million years from now, all the continents will reunite into a single large supercontinent called Neopangaea. Then the process of continental breakup and drift will begin again, forming new continents and oceans that will bear no resemblance to our present-day world.

5

SEAFLOOR SPREADING

The bottom of the ocean was once thought to be barren and featureless, covered by thick, muddy sediments washed off the continents and by debris from dead marine organisms, piled up several miles thick after billions of years of accumulation. During the mid-1800s, soundings were made of the ocean floor before laying the first transcontinental telegraph cable, linking the United States with Europe. The depth recordings indicated hills and valleys and a mid-Atlantic rise, named Telegraph Plateau, where the ocean was thought to be the deepest.

In 1874, the British cable-laying ship HMS *Faraday* was attempting to mend a broken telegraph cable in the North Atlantic. The cable had broken at a depth of 2.5 miles, where it passed over a large rise in the ocean floor. While grappling for the broken cable, the ship caught the strong claws of its grapnel on a rock. With the winch straining to free the grapnel, it finally came loose and was brought to the surface. Clutched in one of its claws was a large chunk of black basalt, a volcanic rock found where volcanoes should not have been.

THE MID-ATLANTIC RIDGE

In 1872, the British corvette HMS *Challenger*, the world's first fully equipped oceanographic vessel, was commissioned to explore the oceans.

Scientists took soundings, water samples, temperature readings, and dredged bottom sediments for evidence of animal life living on the deep ocean floor. Hundreds of species that had never been seen before were brought to the surface. After nearly four years, the research ship charted 140 square miles of ocean bottom and sounded every ocean except the Arctic. The deepest sounding was taken off the Mariana Islands in the Pacific from a depth of 5 miles.

In the early days of sampling sediments on the ocean floor, scientists used a dredge, which was like a bucket tied to the end of a steel cable. The major problem with this technique was that it sampled only the topmost layers of the ocean floor, which could not be recovered in the order they were laid down. In the early 1940s, scientists in Sweden invented a piston corer designed to retrieve a vertical section of the ocean floor intact. The piston corer consisted of a long barrow that plunged into the bottom mud

Figure 56   Piston coring in the Gulf of Alaska. Photo 1 by P. R. Carlson, courtesy USGS

under its own weight (Figures 56 & 57). A piston was fired upward from the lower end of the barrow, sucking up sediments into a pipe. Scientists were thus able to bring up long, cylindrical cores of the ocean floor that dated millions of years old.

The oceanographic research vessel *Glomar Challenger*, commissioned for the Deep Sea Drilling Project in 1968, was developed by a consortium of American

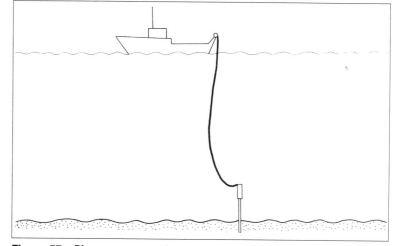

Figure 57   Piston corer on the ocean floor.

oceanographic institutions. Its primary purpose was to drill the ocean floor and take rotary core samples at hundreds of sites scattered around the world. The drill ship dangled beneath it a string of drill pipe as much as 4 miles long. When the drill bit reached the ocean bottom, it bored through the sediments under its own weight. The core was then retrieved through the drill string and brought to the surface.

After dating several cores taken from around various midocean ridges, the scientists discovered something truly remarkable. The sediments were found to be older and thicker the farther the ship drilled away from the deep-sea ridges. But what was even more surprising was that the thickest and oldest sediments were not billions of years old as expected but less than 200 million years old.

In order to measure these sediments, scientists invented a seismic device for underwater use. Seismic waves, which are similar to sound waves, are used to locate sedimentary structures such as those that trap oil. An explosive charge or an air gun is set off below the sea surface, and the seismic waves are picked up by an array of hydrophones trailing behind a ship. Because seismic waves travel slower in soft sediments than they do in hard rock, scientists could use the data to calculate the thickness of different rock layers.

An ocean-bottom seismograph was also lowered to the ocean floor, where it could record microearthquakes in the oceanic crust and then automatically rise to the surface for recovery (Figure 58). These geophysical methods provided scientists with information about the ocean floor that could not be obtained by direct means. Some of their findings, however, came as a complete surprise. Instead of sediments miles thick, they found on average sediments only a few thousand feet thick. It was as though some sort of vacuum cleaner had swept the sediments off the ocean floor.

Sonar, a device which bounced sound waves off the ocean bottom, gave scientists an important tool for mapping undersea terrain. As ships criss-crossed the Atlantic Ocean, onboard sonographs painted a remarkable picture of the ocean floor. Lying in the middle of the Atlantic Ocean was a huge submarine mountain range, surpassing in scale the Alps and Himalayas combined. But what was truly remarkable about this unusual mountain range was that it bisected the Atlantic almost exactly down the middle, weaving halfway between the continents (Figure 59).

As more detailed maps of the ocean floor were made, the Mid-Atlantic Ridge, as it was later named, became the most peculiar mountain range ever known. Through the middle of the 10,000-foot-high ridge crest ran a deep trough as though it were a giant crack in the Earth's crust. It was 4 miles deep in places, or four times deeper than the Grand Canyon, and up to 15 miles wide, making it the largest canyon in the world.

Figure 58   An ocean bottom seismograph provides direct observations of earthquakes on midocean ridges. Photo courtesy USGS

THE SPREADING SEAFLOOR

In the late 1950s, during the height of the cold war, American and Soviet oceanographic vessels crisscrossed the seas, mapping the ocean floor so that ballistic missile submarines could navigate in deep water without bumping into uncharted seamounts. After all the data had been compiled, the maps showed something that was entirely unexpected. Submerged mountains and undersea ridges formed a continuous 40,000-mile-long chain (Figure 60) that was several hundred miles wide, up to 10,000 feet high, and circled the globe like the stitching on a baseball.

Even though it was deep beneath the sea, this midocean ridge system easily became the dominant feature on the face of the planet, extending over an area greater than all the major terrestrial mountain ranges combined. Moreover, it exhibited many unusual features, including massive peaks, sawtooth ridges, earthquake-fractured cliffs, deep valleys, and lava formations of every description (Figure 61). Along much of its length, the ridge system is carved down the middle by a sharp break or rift that is the center of an intense heat flow. Moreover, the midocean ridges were the sites of frequent earthquakes and volcanic eruptions. It appeared that the entire system was a series of giant cracks in the Earth's crust.

The midocean ridge does not form a continuous line, but is broken into small, straight sections called spreading centers. The axis of the midocean ridge is offset laterally in a roughly east-west direction by transform faults (Figure 62), which range from a few miles to a few hundred miles long, encountered every 20 to 60 miles along the ridge system. The transform faults were created when pieces of oceanic crust slid past each other. They produced a series of fracture zones, which are long, narrow linear regions up to about 40 miles wide, of irregular ridges and valleys aligned in a stairstep fashion.

Several spreading centers, 20 to 30 miles long, are separated by nontrans-form offsets up to 15 miles wide. The end of one spreading center might run past the end of another, or the tips of the segments might bend toward each other. The friction between the plates gives rise to strong shearing forces, wrenching the ocean floor into steep canyons. These faults appeared to result from lateral strain,

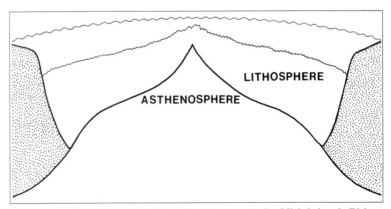

Figure 59   Cross section of the Earth beneath the Mid-Atlantic Ridge.

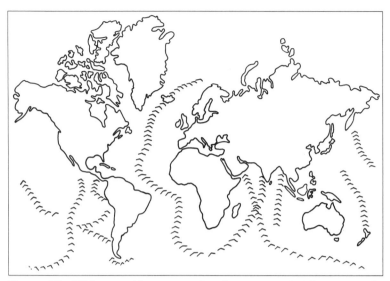

Figure 60   Midocean ridges comprise the most extensive mountain chains in the world.

which is the way movable plates react on the surface of a sphere.

This activity appeared to be more intense in the Atlantic where the mid-ocean ridge is steeper and more jagged, than in the Pacific or the Indian Oceans, where branches of oceanic ridges actually dive under continents. The ocean floor was indeed far more active and younger than it had been imagined. Temperature surveys showed heat seeping out of the Earth in the mountainous regions of the middle Atlantic. Volcanic activity in the ridges indicated that they were constantly adding new material to the ocean floor.

In contrast, the trenches just off the continents suggested that they were sites where old oceanic crust disappeared into the Earth's interior. Gravitational surveys in these regions were highly confusing. In the deep trenches off the edges of certain continents, gravity was found to be much too weak to be responsible for the downward pull on the sediments in vast geosynclines, which are elongated basins in the crust caused by the weight of sediments.

Scientists explained the cracks in the ocean floor by envisioning an expanding Earth due to an increase in the internally generated heat or a decrease in gravity. The

Figure 61   Rim of a lava lake collapse pit on the Juan de Fuca Ridge in the Pacific Ocean. Photo courtesy USGS

weakening of the gravitational field could cause the Earth to bulge out, forming cracks along the crust like cracks on a boiled egg. Most scientists, however, rejected the expansion hypothesis because the force of gravity is considered a constant and has never been known to change. Furthermore, if the Earth of today was significantly larger than in the past, there would be obvious defects in the shapes of the continents, and their continental

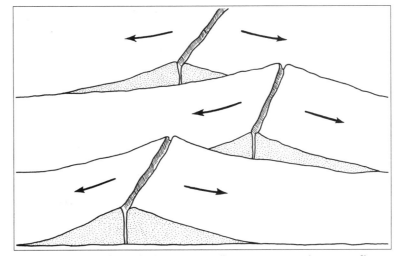

Figure 62   Transform faults at spreading centers on the ocean floor.

margins would not fit as well as they do during reconstructions. Observation of these and other interesting features on the ocean floor led the American geologist Harry Hess to propose a process called seafloor spreading in 1962. His hypothesis described the creation and destruction of the ocean floor, but it did not specify rigid lithospheric plates. Hess came up with a solution that relied on an outward expansion of a different sort, which relied on the Earth's internal convection currents to raise material to the surface.

According to Hess's theory, the process of seafloor spreading first begins with hot rocks rising toward the surface by convection currents in the upper mantle. Upon reaching the underside of the lithosphere, composed of the solid portion of the upper mantle and the overriding crust, the mantle rocks spread out laterally, cool, and descend back into the Earth's interior. The mantle completes a single convection loop in perhaps 100 million years or so. The constant pressure against the bottom of the lithosphere creates fractures, which cause it to rift apart. As the convection currents flow out on either side of the fracture, they carry the now separated parts of the lithosphere along with them, and the opening continues to widen. This reduces the pressure, and the mantle rocks melt and ascend through the fractured zone, where the molten rock finds easy passage through the 60 miles or so of lithosphere until it reaches the bottom of the crust. There, it forms magma chambers that further press against the crust and widen it. In the meantime, magma pours out onto the ocean floor from the trough between ridge crests, adding layer upon layer of basalt to both sides of the spreading ridge.

The pressure of the upwelling magma forces both sides of the ridge farther apart and pushes the ocean floor away from the midocean ridge.

Since any new material added to the ocean floor at midocean ridges must be subtracted somewhere else, Hess suggested that the old seafloor and the lithosphere, upon which it rides, were destroyed in the deep-sea trenches at the edges of continents and along volcanic island arcs. The rocks dive back into the Earth, where they are broken up, remelted, and reabsorbed into the mantle to be used over again in a continuous cycle.

Hess's theory cleared up many problems connected with the mysterious features on the ocean floor, including the midocean ridges, the relatively young ages of rocks in the ocean crust, and the formation of island arcs. But, more importantly, here at last was the long-sought mechanism for continental drift. The continents do not plow through the ocean crust, like an icebreaker slices through frozen seas as Alfred Wegener envisioned, but instead ride like ships frozen in ice floes.

MAGNETIC STRIPES

The more scientists probed the ocean floor, the more complex it became. Studies using sensitive magnetic recording instruments, called magnetometers, towed behind ships over the midocean ridges, revealed the magnetic patterns locked in the volcanic rocks on the ocean floor. The magnetic fields captured in the rocks showed not only the past position of the magnetic poles but their polarity as well.

TABLE 8   COMPARISON OF MAGNETIC REVERSALS WITH OTHER PHENOMENA
(dates are in millions of years)

Magnetic Reversal	Unusual Cold	Meteorite Activity	Dropping Sea Level	Mass Extinctions
0.7	0.7	0.7		
1.9	1.9	1.9		
2	2		2	
10			10	11
40			40	37
70			70	65
130			130	137
160			160	173

SEAFLOOR SPREADING

Recognition of the reversal of the geomagnetic field began in the early 1950s. Over the last 170 million years, the Earth's magnetic field has reversed 300 times. No reversals occurred during the Permian period around 170 million years ago and the Cretaceous period around 90 million years ago. Furthermore, there was a sudden polar shift of 10 to 15 degrees between 100 million and 70 million years ago. Since about 90 million years ago, reversals have steadily become more frequent, and the polar wandering, described above, has decreased to only about 5 degrees.

The last time the geomagnetic field reversed was about 730,000 years ago, and it appears that the Earth is overdue for another one. Two thousand years ago, the magnetic field was considerably stronger than it is today. The Earth's magnetic field seems to have weakened over the past 150 years, amounting to a loss of about 1 percent per decade. If the present rate of decay continues, the field could reach zero and go into another reversal within the next 2000 years.

In 1963, the British geologists Fred Vine and Drummond Mathews thought that magnetic reversal would be a decisive test for seafloor spreading. As the basalts of the midocean ridges cool, the magnetic fields of their iron molecules line up in the same direction as the Earth's magnetic field. When the ocean floor spreads out on both side of the ridge, the basalts record the Earth's magnetic field during each successive reversal like a magnetic tape recording of the history of the geomagnetic field. Normal polarities in the rocks are reinforced by the present magnetic field, whereas reversed polarities are weakened by it. This produces parallel bands of magnetic rocks of varying width and magnitude on both sides of the ridge that are mirror images of each other (Figure 63).

Magnetic reversals also provided a means of dating practically the entire ocean floor because the reversals occur randomly, and any set of patterns are unique in geologic history. The Canadian geophysicist J. Tuzo Wilson calculated the age of a number of magnetic stripes in selected parts of the ocean floor. Calculating the rate of spreading was then a simple manner of determining the age of the magnetic stripes by radiometric dating techniques and measuring the distance from their points of origin at the midocean ridges. The rate of seafloor spreading has changed little over the past 100 million years. Periods of increased acceleration in the past, however, have

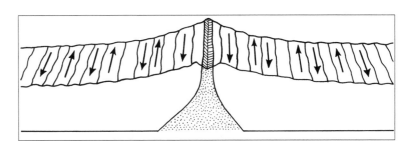

Figure 63   As volcanic rock cools at midocean ridges, it is polarized in the direction of the Earth's magnetic field, producing a series of magnetic stripes on the ocean floor.

Figure 64   Mauna Loa Volcano on the main island of Hawaii is the world's largest shield volcano.
Photo courtesy USGS

been accompanied by an increase in volcanic activity. During the past 10 to 20 million years, there was progressive acceleration, reaching a peak about 2 million years ago.

In the Pacific Ocean, the spreading rates are upward of 6 inches per year, which accounts for less topographical relief at midocean ridges; the active tectonic zone of a fast-spreading ridge is usually quite narrow, generally less than 4 miles across. In the Atlantic the rates are much slower, only about 1 inch per year, which allows taller ridges to form. The Atlantic rift appeared to have opened up sometime between 200 and 150 million years ago. This is remarkably concurrent with Wegener's estimates for the breakup of the continents and the age of the oldest part of the ocean floor, which globally averages only 100 million years old.

SEAMOUNTS AND GUYOTS

The two types of volcanic eruptions associated with midocean ridges are fissure eruptions, the most common type, and those that build typical conical volcanic structures. During fissure eruptions, the magma oozes onto the ocean floor in the form of lava that bleeds through fissures in the trough between ridge crests and along lateral faults. Magma welling up along the entire length of the fissure forms large lava pools, similar to those of broad shield volcanoes such as the Hawaiian volcano Mauna Loa, the largest of its kind in the world (Figure 64).

The two main types of lava formations in the midocean ridges are sheet flows and pillow, or tube, flows. Sheet flows are prevalent in the active volcanic zone of fast-spreading ridge segments, such as those of the East Pacific Rise, and consist of flat slabs of basalt usually less than 8 inches thick. The lava that forms sheet flows is much more fluid than that responsible for pillow formations.

Pillow lavas (Figure 65) appear as though basalt were squeezed out of a giant toothpaste tube. They are mostly found in slow-spreading ridges such as the Mid-Atlantic Ridge, where the lava is much more viscous. The surface of the pillows often has corrugations or small ridges pointing in the direction of flow. The pillow lavas typically form small, elongated hills, pointing downslope from the crest of the ridge. Below the pillow lavas is a middle layer composed of a sheeted-dike complex, which resembles a deck of cards standing on end. Within this structure is a tangled mass of feeders, which bring magma to the surface. Below this is a layer composed of gabbros, which are coarse-grained basalts that crystallized slowly under high pressure.

Upwelling magma from the upper mantle at depths of more than 60 miles below the surface is concentrated in comparatively narrow conduits that lead to the main feeder column. When erupted on the ocean floor, this forms elevated volcanic structures called seamounts, which are isolated and generally strung out in chains across the interior of a plate. Some seamounts are associated

Figure 65 Pillow lava on the south bank of Webber Creek, Eagle District, Alaska. Photo 353 by E. Blackwelder, courtesy USGS

with extended fissures, along which magma has welled up through a main conduit, piling successive lava flows on top of one another.

Seamounts associated with midocean ridges that grow high enough to break the surface of the ocean become volcanic islands. Undersea volcanoes called guyots (pronounced "ghee-ohs") located in the Pacific once towered above the sea. However, constant wave action eroded them below the sea surface as though the tops of the cones were sawed off (Figure 66). The remarkable thing about these volcanoes is that the farther away they were from volcanically active regions of the ocean, the older and flatter they became. This suggests that the guyots wandered across the ocean floor away from their places of origin. In this respect, islands appear to have been produced assembly-line fashion, each moving away in succession from a magma chamber called a hot spot lying beneath the ocean crust.

The most prominent island chain is the Hawaiian Islands, where the youngest and most volcanically active island is Hawaii to the southeast, with progressively older islands having extinct volcanoes, trailing off to the northwest. Continuing from there, coral atolls and shoals were formed when successive layers of coral grew on the flattened tops of long extinct volcanoes that were worn down below sea level. From these islands extend an associated chain of undersea volcanoes known as the Emperor Seamounts. The Hawaiian Islands also lie parallel to two other island chains, the Austral Ridge and the Tuamoto Ridge (Figure 67). The volcanic islands associated with the Mid-Atlantic Ridge system include Iceland, the Azores, the Canary and Cape Verde Islands off West Africa, Ascension Island, and Tristan da Cunha. The volcanic islands associated with the East Pacific Rise are the Galapagos Islands west of Ecuador.

Volcanoes formed on or near the midocean ridges can develop into isolated peaks as they move away from the ridge axis during seafloor spreading. The ocean floor thickens as it leaves the midocean ridge. This process can influence a volcano's height, because, as it moves away from the spreading ridge axis, the thicker crust can support a greater mass on the ocean floor. The ocean crust can also bend like a rubber mat under the massive weight of a seamount. For instance, the crust beneath Hawaii bulges as much as 6 miles. A volcano formed at a midocean ridge cannot increase its mass unless it continues to be supplied with magma after it leaves the vicinity of the ridge. When its source of magma is cut off, erosion begins to

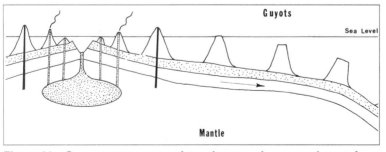

Figure 66 Guyots were once active volcanoes that moved away from their magma source and have since disappeared beneath the sea.

wear it down until it fi-
nally sinks below the
sea.

If a midocean ridge
passes over a hot spot,
the plume directly under
the spreading center aug-
ments the flow of molten
rock welling up from the
asthenosphere to form
new crust (Figure 68).
The crust over the hot
spot is therefore thicker
than it is along the rest of
the ridge, resulting in a
plateau rising above the
surrounding seafloor.
The Ninety East Ridge is
a ruler-straight undersea
mountain chain that runs
300 miles south of the
Bay of Bengal, India, and

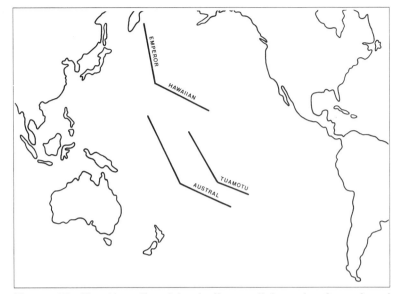

Figure 67 The Hawaiian Islands lie parallel to the Austral and Tuamoto Ridges, indicating the direction of plate travel.

was formed when the Indian plate passed over a hot spot on its way to Asia.

The most striking example of this type of hot-spot volcanism is Iceland (Figure 69), which straddles the Mid-Atlantic Ridge and was raised above the sea about 16 million years ago. Along the ridge, the abnormally elevated topography extends in either direction for about 900 miles, of which 350 miles lies above sea level. South of Iceland, the broad plateau tapers off to form the typical Mid-Atlantic Ridge. The powerful upwelling currents also produce glacier-covered volcanic peaks up to 1 mile high. In 1918, an eruption under a glacier unleashed a flood of meltwater that was 20 times greater than the flow of the Amazon, the world's largest river. In a geologically brief period, the Mid-Atlantic Ridge will move away from the hot spot, carrying Iceland along with it. Devoid of a source of magma, the previously active volcanoes will cease erupting, and Iceland will become just a cold ice-covered, uninhabitable island.

Most volcanoes never make it to the surface of an ocean and remain as isolated undersea volcanoes. Since the crust under the Pacific Ocean is more volcanically active, it has a higher density of seamounts than the Atlantic or Indian Oceans. The number of undersea volcanoes increases with increasing age and thickness of the crust. The tallest seamounts, which rise over 2.5 miles above the seafloor, are located in the western Pacific near the Philippine Islands, where the crust is more than 100

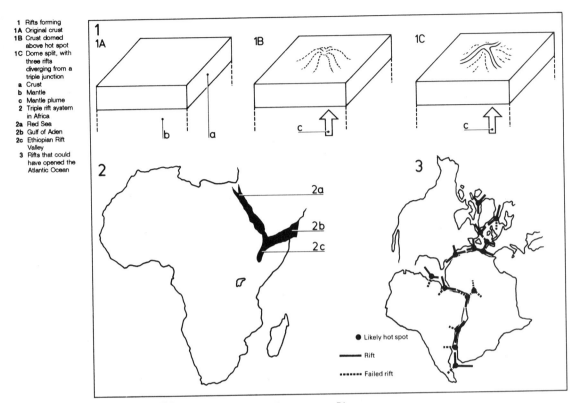

1 Rifts forming
1A Original crust
1B Crust domed above hot spot
1C Dome split, with three rifts diverging from a triple junction
a Crust
b Mantle
c Mantle plume
2 Triple rift system in Africa
2a Red Sea
2b Gulf of Aden
2c Ethiopian Rift Valley
3 Rifts that could have opened the Atlantic Ocean

● Likely hot spot
—— Rift
▪▪▪▪▪▪ Failed rift

Figure 68   Rifts and the rifting process. Illustration courtesy Diagram

million years old. The average density of Pacific seamounts is between 5 and 10 volcanoes per 5000 square miles of ocean floor, a considerably higher density than on the continents.

ABYSSAL RED CLAY

Further evidence for seafloor spreading was found with the discovery of abyssal red clay. The floor of the Atlantic is like a huge conveyer belt, transporting lithosphere from its point of origin at the Mid-Atlantic Ridge to its final destination down the Pacific subduction zones. The ocean floor at the crest of the midocean ridge consists mostly of basalt. Continuing away from the crest, the bare rock is covered with an increasing thickness of sediments, composed mostly of red clay from detritus material washed off the continents. Near the ridge crest, the sediments are predominantly composed of calcareous ooze built up by a rain of decomposed skeletons from microorganisms.

Farther away from the ridge crest, the slope falls below the calcium carbonate compensation zone, where calcareous sediments dissolve in seawater at an average depth of about two miles. Therefore, only red clay should exist in deep water far from the crest of the midocean ridge. However, rock cores taken from the abyssal plains near continental shelves off the edges of continents, where the oceanic crust is the oldest and the deepest, indicated thin layers of calcium carbonate below thick beds of red clay and above hard volcanic rock. Apparently, the red clay protected the calcium carbonate from being dissolved in the deep waters of the abyssal. This implies that the midocean ridge was the source of the calcium carbonate discovered near continental margins and that the seafloor appeared to be moving across the ocean basin, carrying the sediments along with it.

Figure 69   Seawater is sprayed on lava flow in the outer harbor of Vestmannaeyjar, Iceland, from the May 1973 eruption of Heimaey. Photo courtesy USGS

6

SUBDUCTION ZONES

According to the theory of plate tectonics, the *creation* of new oceanic crust at midocean ridges is matched by the *destruction* of old oceanic crust at subduction zones, where lithospheric plates along with their overlying sediments are forced down into the mantle. The rate of seafloor spreading is not always the same as the rate of subduction, however. This results in a lateral motion of the associated midocean ridges. Most of the subduction zones are in the western Pacific, which explains why nowhere is the oceanic crust older than about 170 million years.

The oceanic crust is remarkable for its consistent thickness and temperature, averaging about 4 miles thick and not varying more than 20 degrees Celsius over most of the globe. As the rigid lithospheric plate carrying the oceanic crust descends into the Earth's interior, it slowly breaks up and melts. Over millions of years, it is absorbed into the general circulation of the mantle. The subducted plate also supplies molten magma for volcanoes, most of which ring the Pacific Ocean in a zone called the ring of fire.

DEEP-SEA TRENCHES

In the early 1870s, the British oceanographic ship HMS *Challenger* was taking samples in the deep waters of the Mariana Islands in the western

Pacific when it encountered a deep trough known as the Marianas Trench. It forms a long line northward from the Island of Guam and is the deepest trench in the world, reaching a depth of about 36,000 feet below sea level.

Since the earth is not expanding as it was once thought, new ocean crust created at midocean ridges had to be displaced somewhere else, and the deep-sea trenches seemed to be the most likely places. It had long been thought that the trenches (Figure 70) were great bulges in the ocean crust called geosynclines, resulting from the tremendous weight of sediments washed off the continents and pulled down into the mantle by the weight of a dense underlying material. Gravity surveys conducted over the trenches, however, indicated that no such material existed. It was found that the gravity in the area was much too weak to be responsible for the downward pulling of the seafloor.

It was generally thought that the push plates received from the expansion of ocean floor would be sufficient to force them into the mantle at subduction zones. However, drag at the base of the plates can greatly resist plate motion. Therefore, an additional source of energy is needed to drive the plates. For this purpose, the force of gravity is called upon to provide the driving mechanism, and pull is favored over push to overcome the resistance caused by plate drag.

The farther a plate extends from its place of origin at a midocean spreading center, the thicker and colder it becomes. By the time the plate reaches a subduction zone, it has cooled so much since its formation that it begins to thicken as more material from the asthenosphere adheres to its underside. Eventually, the plate becomes so dense it sinks into the mantle, and the line of subduction creates a deep-sea trench. As the subducted portion of the plate dives into the mantle, the rest of the plate, which might be carrying a continent, is pulled along with it.

An additional force that might help to overcome the resistance caused by plate drag is the pull the sinking plate

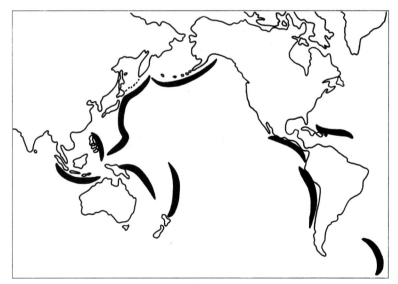

Figure 70 The major trenches of the world, where crustal plates are subducted into the Earth's interior.

receives by mantle convection currents. The magnitude of this force depends on the length of the subduction zone, the rate of subduction, and the amount of trench suction. With these forces in place, the plates could practically drive themselves without the aid of seafloor spreading. Therefore, the upwelling of magma at midocean ridges might be simply a passive response to the plates being pulled apart by subduction.

SEAFLOOR TOPOGRAPHY

The topography of the ocean surface measured by radar altimetry from *Seasat* satellite (Figure 71) shows large bulges and depressions with a relief between bulges and depressions as large as 600 feet. Because these surface variations are spread out over a wide area, they remain undetected by the human eye.

Figure 71   Artist's conception of *Seasat A* satellite as it studies the oceans from Earth orbit.   Photo courtesy NASA

The shape of the surface is dictated by the pull of gravity from undersea mountains, ridges, trenches, and other structures of varying mass distributed over the ocean floor. Therefore, the variations in the height of the ocean surface are influenced by variations in the Earth's gravity field.

Massive undersea mountains produce large gravitational forces, which make the water pile up around them, resulting in swells on the ocean surface. Because submarine trenches have less mass to attract the water, troughs form in the sea surface over these undersea structures. A trench

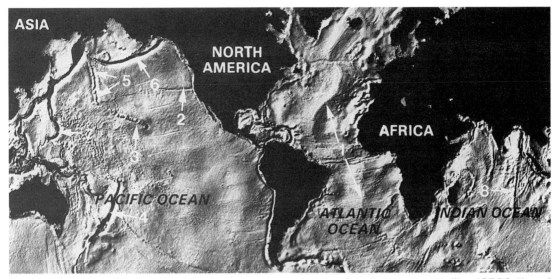

Figure 72 Radar altimeter data from the geodynamic experimental ocean satellite (GEOS-3) and *Seasat* were used to produce this map of the ocean floor. Numbers indicate (1) Mid-Atlantic Ridge; (2) Mendocino fracture zone; (3) Hawaiian Island chains; (4) Tonga Trench; (5) Emperor Seamounts; (6) Aleutian Trench; (7) Marianas Trench; (8) Ninety East Ridge. Photo courtesy NASA

that is 1 mile deep can cause the ocean to drop by dozens of feet. A gravity low, a deviation of the gravity value from the theoretical value, formed as a plate sinks into the mantle off Somalia in northeast Africa, might well be the oldest trench in the world.

The satellite alimetry data was used to produce a map of the entire ocean surface, representing the seafloor as much as 7 miles deep (Figure 72). Chains of midocean ridges and deep-sea trenches were delineated much more clearly than any other method of remote sensing of the ocean floor. The maps uncovered many new features, such as rifts, ridges, seamounts, and fracture zones, and better-defined features already known to exist. The maps also provide support for the theory of plate tectonics, which holds that the Earth is broken into several major plates that shift about, crashing into or moving away from each other. These motions are responsible for all geologic activity taking place on the Earth's surface.

For the first time, geologists were able to view the ancient midocean ridge that formed when South America, Africa, and Antarctica started to separate around 125 million years ago. This particular seafloor spreading center was well concealed due to the buildup of sediment. The boundary between the plates migrated to the west, leaving behind the ancient ridge, which began to subside. Its discovery might help trace the evolution of the oceans and continents during the last 200 million years.

TABLE 9   DIMENSIONS OF DEEP-OCEAN TRENCHES

Trench	Depth (miles)	Width (miles)	Length (miles)
Aleutian	4.8	31	2300
Japan	5.2	62	500
Java	4.7	50	2800
Kuril-Kamchatka	6.5	74	1400
Marianas	6.8	43	1600
Middle America	4.2	25	1700
Peru-Chile	5.0	62	3700
Philippine	6.5	37	870
Puerto Rico	5.2	74	960
South Sandwich	5.2	56	900
Tonga	6.7	34	870

A newly revealed fracture zone in the southern Indian Ocean might shed some light on India's break from Antarctica round 180 million years ago. The 1000-mile-long gash, located southwest of the Kerguelen Islands, was gouged out of the ocean floor as the Indian subcontinent inched northward.

When India collided with Asia, over 100 million years after it was set adrift, it pushed up the Himalaya Mountains to great heights, like squeezing an accordion. A strange series of east- west wrinkles in the ocean crust just south of India verifies that the Indian plate is pushing northward, continuously raising the Himalayas. A sharp bend in the long Mendocino Fracture Zone that juts out of northern California confirms that the Pacific plate abruptly changed direction about the same time India rammed into Asia.

The computer-generated satellite imagery also revealed long-buried parallel fracture zones not seen on conventional seafloor maps. The faint lines running like a comb through the central Pacific seafloor might be the result of convection currents in the mantle 30 to 90 miles beneath the ocean crust. Each circulating loop consists of hot material rising and cooler material sinking back into the depths, tugging on the ocean floor as it descends.

PLATE SUBDUCTION

As a plate moves farther away from its place of origin at a spreading ridge system, it cools and becomes denser as additional mantle material adheres to its underside in a process called underplating. Eventually, the plate becomes so heavy it can no longer remain on the surface and sinks into the mantle, forming a deep sea trench at the point of subduction. The depth at which the oceanic crust sinks as it moves away from the midocean ridges varies with its age. Crust that is 2 million years old lies about 2 miles deep; crust that is 20 million years old lies about 2.5 miles deep; and crust that is 50 million years old lies about 3 miles deep.

The suduction of the lithosphere plays a very significant role in global tectonics and accounts for many of the geologic processes that shape the surface of the Earth. The seaward boundaries of the subduction zones are marked by the deepest trenches in the world and are usually found at the edges of continents or along volcanic island arcs (Figure 73). Major mountain ranges and most volcanoes and earthquakes are associated with the subduction of lithospheric plates.

The amount of subducted plate material is vast. When the Atlantic and Indian Oceans opened up and new oceanic crust was created beginning around 125 million years ago, an equal area of oceanic crust had to disappear into the mantle. This meant that 5 billion cubic miles of crustal and lithospheric material were destroyed. At the present rate of subduction, an area equal to the entire surface of the planet will be consumed by the mantle in the next 160 million years.

When two lithospheric plates converge, generally it is the oceanic plate that is bent and pushed under the thicker, more buoyant continental plate. When oceanic plates collide, the oldest, and thus denser plate dives under the youngest plate (Figure 74). The line of initial subduction is marked by a deep-ocean trench. At first the angle of descent is low, but gradually it steepens to about 45 degrees. At this angle, the rate of vertical descent is less than that of the horizontal motion of the

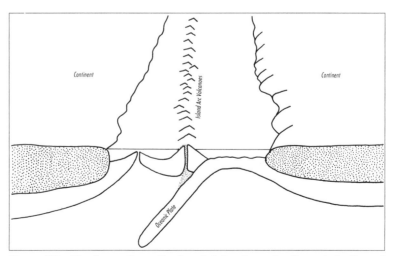

Figure 73   The formation of volcanic island arcs by the subduction of a lithospheric plate.

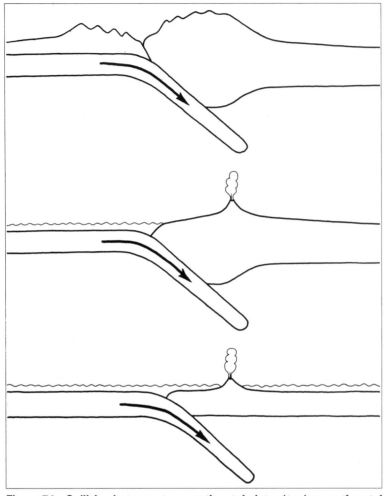

Figure 74   Collision between two continental plates (top), a continental plate and an oceanic plate (middle), and two oceanic plates (bottom).

plate, typically 2 to 3 inches per year.

If continental crust moves into a subduction zone, its lighter weight and greater buoyance prevents it from being dragged down into the trench. When two continent plates converge, the crust is scraped off the subducting plate and plastered onto the overriding plate, welding the two pieces of continental crust together. Meanwhile, the subducted lithospheric plate, now without its overlying crust, continues to dive into the mantle.

In some subduction zones, such as the Lesser Antilles, much of the oceanic sediments and their contained fluids are removed by offscraping and underplating in accretionary prisms that form on the overriding plate adjacent to the trench. In other subduction zones, such as the Marianas and Japan trenches, little or no sediment accretion occurs. Thus, subduction zones differ greatly in the amount of sedimentary material removed at the accretionary prism. In most cases, however, at least some sediment and bound fluids appear to be subducted to deeper levels.

As the continental crust is underthrusted by additional crustal material, the increase in buoyancy pushes up mountain ranges similar to the building of the Himalayas (Figure 75) when India collided with Asia around 40 million years ago. Further compression and deformation might take place beyond the line of collision, producing a high plateau with surface volcanoes, such as the Tibetan Plateau. As resisting forces continue to build up, the plate convergence will eventually stop, the Himalayas will cease growing, and erosion will wear the mountains down.

In the ocean, the deep trenches that are created by descending plates accumulate large deposits of sediments derived from the adjacent continent. The continental shelf and slope, consisting of sediments washed off the continents and the remains of dead marine life, might extend several hundred miles from the edge of the continent. When the sediments and their content of seawater are caught between a subducting oceanic plate and an overriding continental plate, they are subjected to strong deformation, shearing, heating, and metamorphism. The sediments are carried deep into the mantle, where they melt to

Figure 75   View of the Himalaya Mountains of India and China from the space shuttle. Photo courtesy NASA

become the source of new magma for volcanoes (Figure 76). In this manner, the continental crust is rejuvenated, and the total mass of low-density crustal rocks is preserved.

DESCENDING PLATES

When a lithospheric plate descends into the mantle, heat flows into the cooler lithosphere from the surrounding hot mantle. The conductivity of the rocks increases with increasing temperature. Therefore, conductive heating becomes more efficient with depth. Heat of compression is introduced into the plate as it continues to descend and is subjected to increasing pressure. Heat is also generated within the plate by the decay of radioactive elements (mainly uranium, thorium, and potassium), by the change in mineral structure of the rocks, and by internal and external friction, especially at the boundaries between the moving plate and the surrounding mantle. Among these heat sources, conductive heating and friction contribute the most toward heating the descending lithosphere.

At the beginning of descent, the interior of the plate remains relatively cool compared to the mantle. It would at first be subjected to internal stresses, faulting, and fracturing as it bends and dives under another plate. When stresses open a crack in the plate, the weight of the overlying rock layers quickly closes the gap. As the plate begins its passage through the mantle, its temperature rises very slowly due mainly to internal heat sources and more efficient transfer of heat by radiation. Since the heat can no longer escape from the plate, as it did when on the surface, temperatures begin to build up within the plate.

As the plate dives deeper into the mantle, heat is eventually conducted from the outside, where the plate is in contact with the mantle. The plate is also subjected to increasing pressure on its journey through the mantle. By the time the plate reaches a depth of several hundred miles, the extreme temperatures and pressures transform it into a highly dense mineral form with a compact crystal structure. As the plate continues downward, the tightly packed crystals begin to partially melt, and the plate becomes plastic and is able to flow. When the subducted segment of the plate reaches the boundary between the upper and lower mantles, at about 400 miles beneath the surface, it is prevented from descending any farther due to a density difference between the two rock layers, and must bend parallel to the boundary and move in the direction of the convective flow. The entire trip from top to bottom might take some 10 million years. In another 50 million years the plate will have totally lost its identity and be completely assimilated by the mantle.

Not all plates descend even to this level, however. The depth a plate

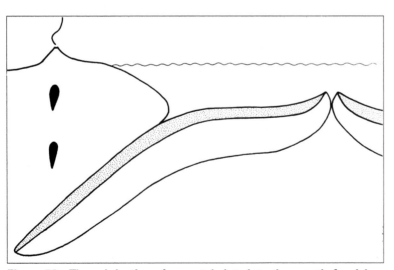

Figure 76 The subduction of a crustal plate into the mantle furnishes volcanoes with molten magma, which rises to the surface in blobs called diapirs.

reaches before being assimilated by the mantle depends on its rate of descent. A slow-moving plate will attain thermal equilibrium at shallower depths. For example, if a plate were descending at a rate of half an inch per year, it would travel no farther than a depth of about 250 miles before complete assimilation occurs. A fast-moving plate will dive deeper into the mantle before it reaches thermal equilibrium.

ISLAND ARCS

Deep-ocean trenches are also regions of intense volcanic activity, and most of the volcanic activity that continually remakes the surface of the Earth takes place at the bottom of the ocean. The trenches were also found to be sites of almost continuous earthquake activity deep in the bowels of the Earth. In 1954, the American seismologist Hugo Benioff discovered a descending lithospheric plate by studying these deep-seated earthquakes, which acted like beacons, marking the boundaries of the plate (Figure 77).

When sediments caught between the subducting oceanic crust and a continental crust are carried deep within the mantle, they are melted in pockets of molten magma called diapirs. When the diapirs reach the underside of the lithosphere, they burn holes in it as they melt their way upward. After reaching the surface, the magma erupts on the ocean floor, sometimes explosively, and in the process creates a new volcanic island or island arc (Figure 78), most of which lie in the Pacific. When cooled, this magma produces a fine-grained, gray rock known as andesite. Named for the Andes Mountains, whose volcanoes were created by a zone of subduction beneath the western portion of the South American plate, andesite is quite different in composition and texture from the upwelling basaltic lavas of spreading midocean rifts. Andesite is indicative of a deep-seated source, possibly as deep as 70 miles within the mantle.

The longest island arc is the Aleutian Islands, which extend more than 3000 miles from Alaska to Asia. The Kurile Islands to the south form another long arc. The islands of Japan, the Philippines, Indonesia, New Hebrides, Tonga, and those from Timor to Sumatra, also form island arcs. The island arcs have similar graceful

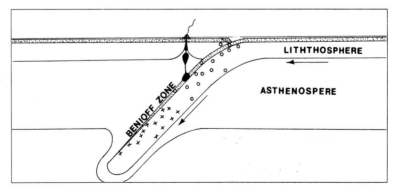

Figure 77   Cross section of a descending lithospheric plate. O's denote shallow earthquakes; X's, deep-seated earthquakes.

Figure 78 A submarine eruption of Myojin-sho Volcano in the Izu Islands, Japan on September 23, 1952. Photo courtesy USGS

curves, and each is associated with a subduction zone. The curvature of the island arcs is due to the curvature of the Earth; when a plane cuts a sphere, the result is an arc.

Behind each island arc lies a marginal or back-arc basin, which is a depression in the ocean crust due to the effects of plate subduction. Steep subduction zones like the Marianas form back-arc basins, and shallow ones like the Chilean subduction zone do not. One classic back-arc basin is the Sea of Japan between China and the Japanese archipelago. Back-arc basins are associated with high heat flow because they are underlain by relatively hot material brought up by convection currents behind the island arcs or by upwelling from deeper regions in the mantle. The trenches have low heat flow because of the subduction of cool dense lithospheric plates, whereas their associated island arcs generally have high heat flow due to a high degree of volcanism.

THE RING OF FIRE

A ring of subduction zones surrounding the Pacific plate is often called the *ring of fire* because of frequent volcanic activity. The subduction zones have devoured almost all the seafloor since the breakup of Pangaea so that the oldest crust lies in a small patch off southeast Japan and dates about 170 million years old. In

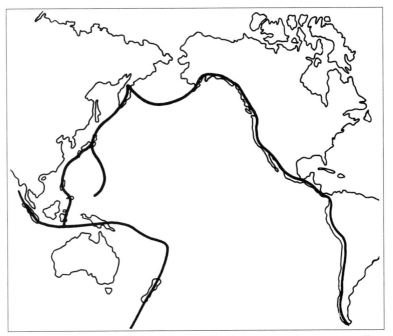

Figure 79 The ring of fire is a ring of subduction zones surrounding the Pacific Ocean.

the process of subduction, the seafloor melts providing molten magma for volcanoes that fringe the deep-sea trenches. Most of the 600 active volcanoes in the world lie in the Pacific Ocean, with nearly half residing in the western Pacific region alone. This produces an almost continuous ring of fire along the edges of the Pacific (Figure 79).

Starting at the western tip of the Aleutian Islands off Alaska, the ring of fire continues along the Aleutian Archipelago and includes the volcanoes of Augustine, Katmai, and Bogoslov. It turns south across the Cascade Range of British Columbia, Washington, Oregon, and northern California. The ring then runs down Baja California and southwest Mexico, where lie the volcanoes of Paricutin and El Chichon (Figure 80).

The band of volcanoes continues down western Central America, which has numerous active cones, including the destructive Nevado del Ruiz of Columbia, which killed 25,000 people in November 1985. The ring of fire journeys along the course of the Andes Mountains on the western edge of South America. It then turns toward Antarctica and the islands of New Zealand, New Guinea, Indonesia, the Philippines, where the June 1991 eruption of Pinatubo was the most powerful eruption of the century, and Japan, finally ending on the Kamchatka Peninsula in eastern Siberia.

The ring of fire coincides with the circum-Pacific earthquake belt because the same tectonic forces that produce earthquakes are also responsible for volcanic activity. The area of greatest seismicity occurs on plate boundaries associated with deep trenches along volcanic island arcs and along the margins of continents. As the Pacific plate inches northwestward, its leading edge dives into the mantle, forming some of the deepest trenches in the world.

By studying a band of earthquakes, scientists are able to see the earliest stages in the birth of a subduction zone that will eventually form a trench to the north and west of New Guinea in the western Pacific. They discovered that gravity was lower than normal, which would be expected over a trench. Furthermore, there is a bulge in the crust to the south, indicating that the edge of a slab of crust is beginning to be thrust into the Earth. If subduction is just being initiated, it will take 5 to 10 million years before the process is fully underway.

The entire western seaboard of South America is also affected by an immense subduction zone just off the coast. The lithospheric plate that the

Figure 80   The El Chichon Volcano after its 1982 explosive eruption in southern Mexico. Photo courtesy USGS

South American continent rides on is forcing the Nazca plate to buckle, causing great tensions to be built up deep within the crust. While some rocks are being forced deep down, others are buoyed up, raising the Andean Mountain chain. The resulting forces are building great stresses into the entire region. When the stresses become large enough, powerful earthquakes roll across the landscape.

Subduction zone volcanoes such as those in the western Pacific and Indonesia (Figure 81) are among the most explosive in the world, creating new islands and destroying old ones, including the near total destruction of the Indonesian island of Krakatoa in 1883. The reason for their explosive nature is that the magma contains large amounts of silica and volatiles, consisting of water and gasses derived from sediments on the ocean floor, that are subducted into the mantle and melted. When the pressure is lifted as the magma reaches the surface, these volatiles are released explosively, fracturing the magma which then shoots out of the volcano like pellets from a shotgun.

The Cascade Range in the western United States is composed of a chain of volcanoes associated with a subduction zone under the North American continent, created when the North American plate overran the northern part of the now extinct Pacific rift system. As the lithosphere is being forced into the mantle, the tremendous heat melts parts of the descending plate and the adjacent lithospheric plate to form pockets of magma. The magma forces its way to the surface, resulting in explosive volcanic eruptions. The May 18, 1980, eruption of Mount St. Helens, which devastated 200 square miles of national forest (Figure 82) is a prime example of the explosive nature of subduction zone volcanoes.

As the Earth ages, the lithosphere will continue to thicken and the asthenosphere will become more viscous, due to the cooling of the mantle after most of the radioactive heat sources have decayed into stable elements. As the plates continue to thicken and activity in the asthenosphere slows down, the movement of the plates becomes sluggish. Eventually they will come to a complete halt. When this happens, perhaps some 2 billion years from now,

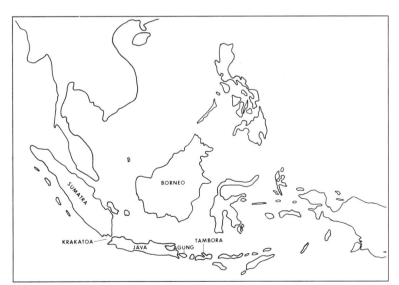

Figure 81   The locations of the great Indonesian volcanoes.

Figure 82   An entire forest was leveled by the explosive eruption of Mount St. Helens.   Photo by Jim Hughes, courtesy USDA–Forest Service

volcanoes will no longer erupt, earthquakes will cease, and the Earth will become a dead planet.

7

MOUNTAIN BUILDING

The most impressive landforms the Earth has to offer are its mountain ranges, created by the forces of uplift and erosion. A mountain is defined as a topographic feature that rises abruptly above the surrounding terrain and involves the massive deformation of the rocks that form the core of the mountain. The usual means of building mountains is for plate motions to shove crust onto another strong plate or create a deep root of light crustal rock that literally floats the mountain like an iceberg.

Most mountains occur in ranges, and although a few isolated peaks do exist, they are rare. Mountains have complex internal structures formed by folding, faulting, volcanic activity, igneous intrusion, and metamorphism. Mountain building, which provides the forces necessary for folding and faulting rocks at shallow depths, also supplies the stress forces that strongly distorts rocks at greater depths.

EPISODES OF OROGENY

Many of today's mountain ranges were uplifted by Paleozoic continental collisions (Figure 83). The collision of North America and Europe in the early Paleozoic was responsible for the Caledonian orogeny, or mountain building, which resulted in a mountain belt that extended from southern Wales, spanned across Scotland, and ran northeastward through Scandinavia. In North America, this orogeny built a mountain belt that extended from Alabama through Newfoundland and reached as far west as Wisconsin and Iowa. Vermont still preserves the roots of these ancient mountains, which were shoved upward some 400 million years ago but have since been planed by erosion.

The Appalachians and the Ouachitas in the United States were formed when North America and Africa slammed into each other during the late Paleozoic from 300 to 250 million years ago. The southern Appalachians are underlain by over 10 miles of sedimentary and metamorphic rock that are essentially undeformed, whereas the surface rocks were highly deformed by the collision (Figure 84). This suggests that these mountains were the product of horizontal thrusting, in which crustal material was carried great distances.

Also raised during this period of continental collision was the Mauritanide mountain chain of western Africa, which is characterized by a series of belts similar to the Appalachians, and in a sense the two mountain ranges are mirror images of each other. This mountain-building episode also uplifted the Hercynian mountains in Europe, which extended from England to Ireland and continued through France and Germany. The Ural Mountains were created by a collision between the Siberian and Russian shields. The Transant-

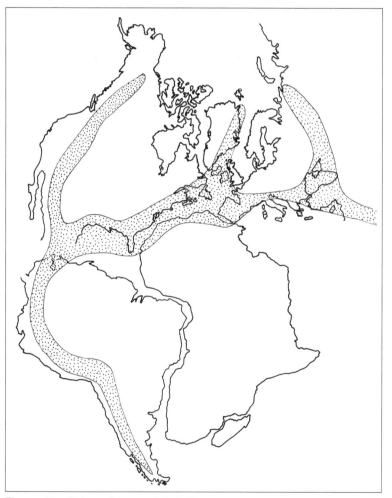

Figure 83   Paleozoic mountain belts.

arctic Range was created when two plates came together to form the continent of Antarctica.

Prior to this time, the supercontinents were separated by the Tethys Sea, and Europe and North America were separated by a proto-Atlantic ocean called the Iapetus Sea. These ancient seas narrowed in the late Paleozoic when the continents rejoined in a *great collision*. Gondwana and Laurasia

TABLE 10   THE TALLEST MOUNTAIN PEAKS BY STATE

State	Mountain or Peak	Elevation (feet)	State	Mountain or Peak	Elevation (feet)
Alabama	Cheaha Mt.	2,407	Montana	Granite Pk.	12,799
Alaska	McKinley Mt.	20,320	Nebraska	Kimball Co.	5,246
Arizona	Humphreys Pk.	12,663	Nevada	Boundary Pk.	13,143
Arkansas	Magazine Mt.	2,753	New Hampshire	Washington Mt.	6,288
California	Whitney Mt.	14,494	New Jersey	High Point	1,803
Colorado	Elbert Mt.	14,433	New Mexico	Wheeler Pk.	13,161
Connecticut	Frissell Mt.	2,380	New York	Marcy Mt.	5,344
Delaware	Ebright Rd.	442	North Carolina	Mitchell	6,684
Florida	Walton Co.	345	North Dakota	White Butte	3,506
Georgia	Brasstown Bald	4,784	Ohio	Campbell Hill	1,550
Hawaii	Mauna Kea Mt.	13,796	Oklahoma	Black Mesa	4,973
Idaho	Borah Pk.	12,662	Oregon	Mt. Hood	11,239
Illinois	Charles Mound	1,235	Pennsylvania	Davis Mt.	3,213
Indiana	Wayne Co.	1,257	Rhode Island	Jerimoth Hill	812
Iowa	Osceola Co.	1,670	South Carolina	Sassafras Mt.	3,560
Kansas	Sunflower Mt.	4,039	South Dakota	Harney Pk.	7,242
Kentucky	Black Mt.	4,145	Tennessee	Clingmans Dome	6,643
Louisiana	Driskill Mt.	535	Texas	Guadalupe Pk.	8,749
Maine	Katahdin Mt.	5,268	Utah	Kings Pk.	13,528
Maryland	Backbone Mt.	3,360	Vermont	Mansfield Mt.	4,393
Massachusetts	Greylock Mt.	3,491	Virginia	Rogers Mt.	5,729
Michigan	Curwood Mt.	1,980	Washington	Mt. Rainer	14,410
Minnesota	Eagle Mt.	2,301	West Virginia	Spruce Knob	4,863
Mississippi	Woodall Mt.	806	Wisconsin	Timms Hill	1,951
Missouri	Taum Sauk Mt.	1,772	Wyoming	Gannett Pk.	13,804

Figure 84 The foothills of Blue Ridge and Piedmont Plateau, Buncombe County, North Carolina. Photo 231 by A. Keith, courtesy USGS

converged into the crescent-shaped supercontinent Pangaea, whose landmass extended almost from pole to pole. A huge ocean stretched uninterrupted across the rest of the planet, while the continents huddled to one side.

The continental *collisions* crumpled the crust, pushing up huge masses of rocks into several mountain chains throughout many parts of the world (Figure 85). Besides folded mountain belts, chains of volcanoes were prevalent as well. During times of highly active continental movements, there is a higher degree of volcanic activity, especially at midocean ridges and subduction zones. The amount of volcanism could affect the rate of mountain building and the composition of the atmosphere, which in turn could affect the climate.

Since the late Cambrian, the future Rocky Mountain region was close to sea level. Farther west within about 400 miles of the coast was a mountain belt comparable to the present Andes. This belt developed between 160 and 80 million years ago above a subduction zone, which might have been responsible for the Cretaceous Sevier orogeny that formed the Overthrust Belt in Utah and Nevada. A region from eastern Utah to the Texas Panhandle was deformed during the late Paleozoic Ancestral Rockies orogeny but was completely eroded before the formation of the present Rocky Mountains. The Rocky Mountain foreland region subsided by as much as 2 miles between 85 million and 65 million years ago. It then rose above sea level and acquired its present elevation around 30 million years ago.

Between 80 million and 40 million years ago, the Larimide orogeny raised the Rocky Mountains, which extended from northern Mexico into Canada. A large part of western North America was uplifted, and the entire Rocky Mountain Region was raised nearly a mile above sea level. The orogeny was caused by an increase in buoyancy of the continental crust, resulting from the subduction of vast areas of oceanic crust and its attached lithosphere beneath the west coast of North America. A region between the Sierra Nevada and the southern Rockies took a spurt of uplift during the past 20 million years, raising the area over 3000 feet. The Canadian Rockies consist of slices of sedimentary rock that were successively detached from the underlying basement rock and thrust eastward on top of one another.

Westward of the Rockies, a large number of parallel faults sliced through the Basin and Range Province between the Sierra Nevada and the Wasatch Mountains, resulting in a series of about 20 peculiar north-south-trending mountain ranges comprised of grabens and horsts. The Basin and Range, which includes southern Oregon, Nevada, western Utah, southeastern California, and southern Arizona and New Mexico, is composed of numerous fault-block mountain ranges bounded by high-angle normal faults. (See Figure 3, page 4.) The crust is literally broken into hundreds of pieces that were steeply tilted and raised nearly a mile above the basin, forming nearly parallel mountain ranges up to 50 miles long. Death Valley (Figure 86), the lowest place on the North American continent at 280 feet below sea level, was once several thousand feet in elevation. The area collapsed when the continental crust thinned due to extensive block faulting in the region. The Great Basin area is a remnant of a broad belt of mountains and high plateaus that subsequently collapsed after the crust was pulled apart following the Laramide. The Andes Mountains of South America could follow a similar evolution sometime in the geologic future.

The still-rising Wasatch Range of north-central Utah and southern Idaho (Figure 87) is an excellent example of a north-trending series of normal faults, one below the other. The fault blocks extend for 80 miles, with a probable net slip along the fault plane on the west side of 18,000 feet. The Tetons of western Wyoming (Figure 88) were upfaulted along the eastern flank and downfaulted to the west. The rest of the Rocky Mountains were created by a similar process of upthrusting connected with plate collision and subduction that raised the Andes Mountains of Central and South America. The Andes Mountains continue to rise due to an increase in crustal buoyancy resulting from the subduction of the Nazca plate beneath the South American plate.

India and the rocks that now make up the Himalayas broke away from Gondwana early in the Cretaceous, sped

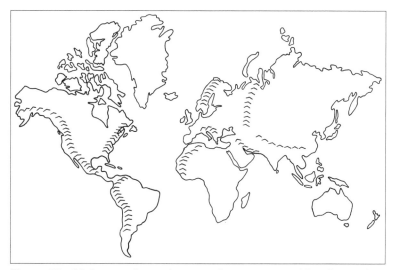

Figure 85   Major continental mountain ranges resulting from plate collisions.

Figure 86   A view southwest across Death Valley, Inyo County, California. Photo 456 by W. B. Hamilton, courtesy USGS

across the ancestral Indian Ocean, and slammed into southern Asia about 40 million years ago. In the process, 6000 miles of subducting plate was destroyed. As the Indian and Asian plates collided, the oceanic lithosphere between them was thrust under Tibet. The increased buoyancy uplifted the Himalaya Mountains and the wide Tibetan Plateau. These regions rose at an incredible rate during the past 5 to ten million years, when the entire area was uplifted more than a mile.

The Tethys Sea separating Eurasia from Africa continued to fill with thick sediments in a vast geosyncline that formed a huge bulge in the Earth's crust. About 50 million years ago, the Tethys narrowed as Africa approached Eurasia, and began to close off entirely some 20 million years ago. Like a rug thrown across a polished floor, the crust crumbled into folds. Thick sediments that had been accumulating for tens of millions of years were compressed into long belts of mountain ranges on the northern and southern continental landmasses.

The entire crusts of both continental plates buckled upward, forming the central portions of the range. This episode of mountain building called the Alpine orogeny, ended about 26 million years ago. It raised the Pyrenees on the border between Spain and France, the Atlas Mountains of northwest

Figure 87   The Wasatch Range of north-central Utah. Photo 305 by R. R. Woolley, courtesy USGS

Africa, and the Carpathians in east-central Europe. The Alps of northern Italy formed in much the same manner as the Himalayas, when the Italian prong of the African plate was thrust into the European plate. But because the European plate is only half as thick as the Indian plate, the Alps are only about half as tall as the Himalayas.

PLATE COLLISIONS

Before the continental drift theory was accepted, there was no adequate explanation for the building of mountain ranges. It was thought that mountains formed early in the Earth's history when the molten crust cooled and solidified, forcing it to contract and shrivel up. However, after further study, geologists were forced to conclude that the folding of rock layers was much too young and intense, and therefore, some other forces had to be at work in raising the world's mountains.

In his theory of continental drift, Alfred Wegener introduced his theory of mountain building. Although it was convincing because of its elegance and simplicity, it was not entirely correct. He thought that as the continents pushed through the ocean floor like an icebreaker plows through arctic ice, the continents encountered increasing resistance, causing the leading edges to crumble, fold back, and thrust upward.

In actual practice, however, when an oceanic plate and a continental plate collide, thousands of feet of sediments are deposited along the seaward margin of the continental plate in deep-ocean trenches, and the increased weight presses downward on the oceanic crust. As the continental and oceanic plates merge, the heavier oceanic plate is subducted or overridden by the lighter continental plate, forcing it further downward. The sedimentary layers of both plates are compressed, resulting in a

Figure 88   The Teton Range, Grand Teton National Park, Teton County, Wyoming. Photo 1260 by I. J. Witkind, courtesy USGS

swelling of the leading edge of the continental crust, forming a mountain belt similar to the Andes of South America (Figure 89).

A gravity survey in the Andes Mountains of South America showed that its gravitational attraction was less than that observed at sea level. This suggested that the granites in the mountains were lighter or less dense than the rocks below and thus exerted less gravitational pull on the instruments. From the data, geologists concluded that the continents were composed of lighter granitelike materials called sial (from *silica* and *aluminum*) and the ocean floor was composed of heavier basaltlike substances called sima (from *silica* and *magnesium*). The difference in density between the two materials made the continents buoyant, a relationship known as *isostasy*, from the Greek word meaning "equal standing." As the oceanic crust descends, the topmost layers are scraped off and plastered against the swollen edge of the continental crust, forming an accretionary prism.

When two continental plates collide they crumple the crust, forcing up mountain ranges at the point of impact (Figure 90). The long-lived continental roots that underlie mountain ranges can extend downward 100 miles or more into the upper mantle. The drifting continents carry along with them thick layers of chemically distinct mantle rock. Apparently through collisions resulting from plate tectonics, continents have stabilized this part of the mobile mantle rock below them. The squeezing of a plate into a thicker one due to continental collision, such as that between the Indian and Asian plates to uplift the Himalayas, might be the very process that forms deep roots.

The sutures joining the landmasses are still visible as the eroded cores of ancient mountains known as orogens. Ancient rocks that make up the interiors of the continent, which were assembled some 2 billion years ago, are called cratons. Caught between the cratons was an assortment of debris swept up by drifting continents, including sediments from continental shelves and the ocean floor, stringers of volcanic rock, and small scraps of continents, all sliced by

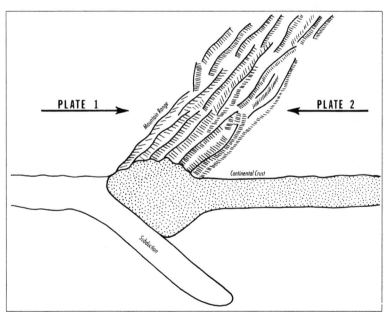

Figure 89 The formation of mountains by the convergence of two crustal plates.

PLATE 1

PLATE 2

Mountain Range

Continental Crust

Subduction

Figure 90 Wax model illustrating mountain uplift. Photo 253 by J. K. Hillers, courtesy USGS

faults. Ophiolites, which are pieces of ocean crust thrust up on land, have been found deep in the interiors of the continents, indicating that they were patched together in the distant past. Blueschists, which are metamorphosed rocks of subducted ocean crust shoved up on the continents, might have also been present at a very early age.

Many major mountain ranges are also associated with the subduction of lithospheric plates. When continental crust moves into a subduction zone, its buoyancy prevents it from being carried downward into the Earth's interior. The crust is scraped off the subducting lithospheric plate and plastered onto the overriding plate. The lithospheric plate, now without its overlying crust, continues to dive into the mantle.

The submerged crust is underthrusted by additional crustal material, and the increased buoyancy raises mountain ranges. Additional compression and deformation might take place farther inland beyond the line of collision, producing a high plateau with surface volcanoes, similar to those on the wide plateau of Tibet, most of which lies at an average elevation of 3 miles above sea level and constitutes one of the largest snowfields in the world.

Knowledge of the complex tectonic underpinning of the Tibetan Plateau is the key to understanding mountain building. Apparently the strain of raising the world's highest mountain range by the collision of the Indian plate with Asia has resulted in deformation and earthquakes all along the plate. India is still plowing into Asia at a rate of about 2 inches a year. As resisting forces continue to build up, the plate convergence will eventually stop, the mountains will cease growing, and crustal weakening and erosion will ultimately bring them down to sea level.

IGNEOUS ACTIVITY IN MOUNTAIN BELTS

Many mountain ranges are associated with volcanic and seismic activity. In the deepest part of the continental crust, where temperatures and pressures are very high, rocks are partially melted and metamorphosed. Pockets of magma also provide the source material for volcanoes and igneous intrusions, which invade the crust to form large bodies of granitic rocks.

Intrusive magma bodies come in a variety of shapes and sizes. The largest intrusives are called batholiths. They are generally greater than 40 square miles in area and are much longer than they are wide.

Figure 91 Kern Canyon in Sequoia National Park, Tulare County, California. Photo 1043 by F. E. Matthes, courtesy USGS

When the overlying sediments are eroded away, the harder, more resistant batholith remains. This results in a major mountain range, such as the Sierra Nevada in California (Figure 91), which extends for nearly 400 miles but is only about 50 miles wide.

Additional buoyancy might have been provided when the underlying lithosphere was replaced by hotter rock, which shoved the mountain range upward. Globs of relatively cold rock from the lithosphere, dropping hundreds of miles into the mantle, appear to precede this type of mountain building. The 2.5-mile-high southern Sierra Nevada has risen some 7000 feet over the last 10 million years, but no plates have converged near the region for more than 70 million years.

Batholiths comprise granitic rocks composed of quartz, feldspar, and mica. The rocks might contain veins, where rich ores have accumulated. The ores were formed when metal-rich fluids from a magma chamber migrated into cracks and fractures in the rocks. Similar to a batholith is a stock, except that a stock is smaller than 40 square miles in area. It may be an extension of a larger batholith buried below. Like batholiths, stocks are also composed of course-grained granitic rocks. Another type of rock formation, a dike, is an intrusive magma body that is tabular in shape and considerably longer than it is wide. It forms when magma fluids occupied a large crack or fissure in the crust. Because dike rocks are usually harder than the surrounding material, they generally form long ridges when exposed by erosion. A sill is similar to a dike in its tabular form except that it is produced parallel to planes of weakness such as sedimentary beds. A special type of sill is called a laccolith, which tends to bulge the overlying sediments upward, sometimes forming mountains like the Henry Mountains in southern Utah (Figure 92).

If magma extrudes onto the Earth's surface either by a fissure eruption, the most common type, or a vent eruption, the volcanism builds majestic mountains. The volcanoes of the Cascade Range (Figure 93) from northern California into Canada resulted from the subduction of the Juan de Fuca plate along the Cascadia subduction zone beneath the northwestern United States. As the plate melts diving into the mantle, it feeds molten rock to magma chambers that underlie the volcanoes. Besides supplying magma for these very powerful volcanoes, the subducting plate also has the potential of generating very strong earthquakes in the region.

EROSION

No matter how pervasive is the formation of mountain ranges by the convergence of crustal plates, erosion is an equally powerful force in wearing them down. Erosion has leveled the tallest mountains, gouged deep ravines into the hardest rock, and obliterated most geologic structures on Earth, including almost all large meteorite-impact structures, which explains why few are still visible.

The rise of active mountain chains such as the Himalayas is matched by erosion so that their net growth is nearly zero. The world's mountain ranges contain some of the oldest rocks. What was once deep below the surface is thrust high above as huge blocks of granite are pushed up by tectonic forces deep within the Earth and exposed by erosion. The process of erosion is delicately balanced by the forces of buoyancy, which keeps the crust afloat. When erosion has shaved off the top 2.5 miles of continental crust, the mean height of the crust falls below sea level, at which point erosion ceases and sedimentation begins as the sea inundates the land and sediments accumulate on the seafloor.

Before the spread of vegetation on the land, soil eroded easily because there were no plant roots

Figure 92   Mount Hillers, Henry Mountains, Garfield County,Utah.
Photo 25 by R. G. Luedke, courtesy USGS

to hold it in place. Therefore, erosion rates were probably much higher than they are now. In the early stages, the relief of the land was not nearly as great as it is today. It took eons of mountain building and erosion to give us our present landscape of tall mountains and deep canyons. This includes the 29,000-foot Mount Everest in the Himalayas, the world's tallest

Figure 93   Mount Baker and Mazama Glacier, Whatcom County, Washington. Photo courtesy USGS

peak, and the Dead Sea in the Middle East, which at 1300 feet below sea level is the lowest point on the Earth's surface.

The Mesozoic era is known for its erosion of previously formed mountain ranges such as the Appalachians, which were eroded to stumps. Large inland lakes formed, and the seas invaded the interiors of most continents. Much of Russia, western Siberia, and Asia were underwater, along with portions of western North America and parts of South America.

Large amounts of sediments accumulated in the inundated areas, and the central seas were crowded out as these deposits grew thicker. By the end of the Mesozoic, salt lakes and swamps were formed by the receding waters. The Mesozoic was also noted for its widespread basalt lava flows, with the most extensive activity in the Southern Hemisphere. These flows were accompanied by large-scale intrusions of igneous rocks, some of which provide our present-day world with much of its mineral wealth.

MINERAL DEPOSITS

Throughout geologic history, in many parts of the world, a piece of crust was scraped off an oceanic plate as it plunged under the leading edge of a continent. These ophiolites, from the Greek *ophis*, meaning "snake or serpent," are associated with igneous rock known as serpentinite, which has a mottled green appearance like that of a serpent.

When former volcanically active regions of the ocean crust are uplifted onto the continents, they provide rich metal ore deposits that are mined throughout the world. The Troodos ophiolite on Cyprus was mined extensively by the early Greeks for its copper and tin. Mining tools dating from before 2500 B.C. have been found in the underground mine workings. The ore provided some of the first bronze for the earliest Greek sculptures. In effect, the Greeks were mining the ocean floor, which had been conveniently brought up to the surface.

As the oceanic crust moves away from a spreading center, it eventually reaches the margin of the ocean basin, where it is either subducted into the mantle or collides head-on with another lithospheric plate and raises mountains. In the course of these events, fragments of oceanic crust might be uplifted and exposed on land. These fragments of ancient oceanic lithosphere have been identified in various parts of the world (Figure 94). The ophiolites consist of an upper layer of marine sediments, a layer of pillow lava (basalts that have erupted undersea), and a layer of dark, dense ultramafic rocks (rich in iron and magnesium and poor in silica) that are thought to be part of the upper mantle. The metal ore deposits are at the base of the sedimentary layer just above the area where it makes contact with the basalt. Examples include ophiolite complexes exposed on the

Apennines of northern Italy, the northern margins of the Himalayas in southern Tibet, the Ural Mountains in the Soviet Union, the eastern Mediterranean including Cyprus, the Afar Desert of northeastern Africa, the Andes of South America, islands of the western Pacific such as the Philippines, uppermost Newfoundland, and Point Sol along the Big Sur coast of central California.

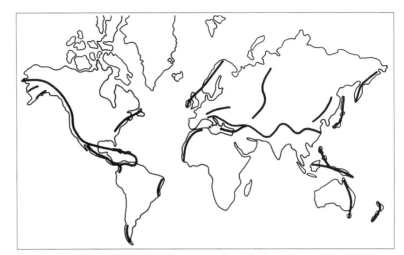

Figure 94   Worlfwide distributuion of ophiolites.

Another type of metal ore deposit formed at oceanic spreading centers is called a massive sulfide. These deposits contain sulfides of iron, copper, lead, and zinc. They occur in most ophiolite complexes, such as the Apennine ophiolites, first mined by the ancient Romans. Massive sulfide deposits are mined extensively in other parts of the world for their rich ores of copper, lead, and zinc.

The theory of plate tectonics provides many clues as to why certain ore bodies are in their present locations. Hydrothermal activity, the movement of hot water in the crust, is a reflection of high heat flow, and high heat flow is associated with plate boundaries. The Red Sea between Africa and Arabia and the Salton Sea of southern California shows evidence of recently transported metals at extensional plate boundaries, where plates are being pulled apart.

In the Salton Sea, the crust is thinning as blobs of magma rise toward the surface, resulting in a hot, fractured crust, whose salt-laden waters are tapped for their geothermal energy. The brine, which is eight times saltier than seawater, also dissolves metals in the sediments. The metals precipitate out of solution and are concentrated in fractures in the rock. The mineralization is similar to that found in rifts that opened more than 600 million years ago and are now mined for their copper, lead, and zinc. The Salton Sea area is still quite young, only 100,000 years old, and its deposits have not yet achieved ore grade. However, given another half million years, it will yield significant ores. Another area of high heat flow is associated with igneous activity at converging plate margins, where one plate is forced under another in subduction zones. Many hydrothermal ore deposits have been found in both new and ancient convergent margins, and doubtless many more future ore deposits will be found.

Plate tectonics has also aided in the exploration for oil and gas. The theory helps to explain why oil reserves are located where they are and might suggest new sites to explore for oil. It is commonly believed that oil and natural gas were formed from the remains of abundant plant and animal life that lived in tropical regions tens of millions to hundreds of millions of years ago. This is true for coal as well, and plant fossils are actually found between coal layers, indicating their organic origin.

Deep burial and heat provided by the Earth's interior created a gigantic pressure cooker, which through geologic time has baked the organic compounds into hydrocarbons (Figure 95). The eventual drifting of the continents and erosion of the surface layers has brought the oil and gas deposits within reach of the driller's bit.

GEOTHERMAL ENERGY

Much of the young mountain terrain in the western United States, as well as in Alaska and Hawaii, is of volcanic origin and forms a tempting but well-locked treasure of geothermal energy, which can provide high-pressure steam for generating electricity. The potential geothermal energy resources in the United States alone is estimated to equal twice the energy of the world's petroleum reserves. The dry hot rock resources are several thousand times greater. Dry hot rocks lie beneath the surface in areas where the thermal gradients are two to three times greater than normal, about 100 degrees Celsius per mile of depth. The process of artificially making a geothermal reservoir within hot buried rocks is difficult and expensive, but if successful, the potential is enormous. Just a single eruption of Kilauea on the main island of Hawaii could supply two fifths of the power requirements of the entire United States during the time of the eruption.

In a sense, the Earth's interior can be thought of as a natural nuclear power reactor because the heat is mainly derived by the decay of radioactive elements. Many steam and geyser areas around the world are generally associated with active volcanism lo-

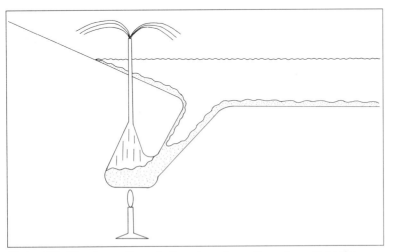

Figure 95   The formation of oil and gas in sedimentary deposits.

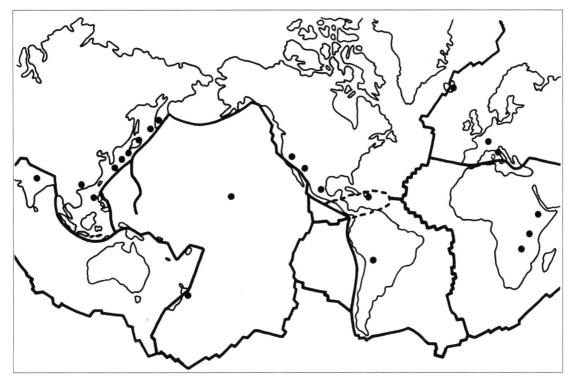

Figure 96 Most of the world's geothermal production is associated with plate boundaries.

cated at plate margins (Figure 96). These are potential sites for tapping geothermal energy for steam heat and electric power generation. Iceland, Italy, Mexico, New Zealand, the Soviet Union, and the United States utilize underground supplies of superheated steam to drive turbine generators for electric power production.

Geothermal energy could prove to be far more valuable in the long run than petroleum, coal, or even nuclear energy. Besides, it is nonpolluting. The Earth's internal heat will last for billions of years. Unlike limited resources of fossil fuels, geothermal energy has the potential of supplying our energy needs for millennia.

8

THE ROCK CYCLE

The development of the theory of plate tectonics has also led to a greater understanding of the geochemical carbon cycle, or simply rock cycle, that is extremely crucial for keeping our planet alive in the biologic sense. The recycling of carbon through the geosphere makes the Earth unique among planets. This is evidenced by the fact that the atmosphere contains large amounts of oxygen, which without the carbon cycle would have long since been buried in the geologic column composed of layers of strata. Fortunately, plants replenish oxygen by utilizing carbon dioxide, which plays a critical role as a primary source of carbon for photosynthesis and therefore provides the basis for all life.

Presently, carbon dioxide makes up about 350 parts per million of the atmosphere, amounting to about 700 billion tons of carbon. It is one of the most important greenhouse gases, whose purpose is to trap solar heat that would otherwise escape into space. In this respect, carbon dioxide operates like a thermostat that regulates the temperature of the planet. If too much carbon dioxide is removed from the atmosphere by the carbon cycle, the Earth would cool down. If too much carbon dioxide is generated by volcanic activity, the Earth would heat up. Therefore, any changes in the carbon cycle could have profound effects on the climate and, ultimately, on life.

THE ATMOSPHERE AND OCEAN

During the early formation of the Earth, volcanoes spewed out massive quantities of gases and steam, during what is known as the "big burp." Volcanoes are responsible for a variety of products, including water vapor, nitrogen, carbon dioxide, methane, ammonia, sulfur dioxide, and a host of other gasses (Figure 97). Water and carbon dioxide are especially abundant in magma, which helps make it flow easily.

Early in the Earth's history, volcanoes erupted much the same as they do today, only on a greater scale. They were also more violent due to the higher temperature of the Earth's interior and the larger amounts of volatiles in the magma, which made the eruptions highly explosive. The early volcanoes were gigantic by today's standards and shot rock fragments and ash 100 miles or more into space. Since the Earth had no atmosphere to scatter the volcanic debris, it simply fell back around the volcanic vent, building up volcanoes to tremendous heights. Some volcanoes were so massive they often could not support their own weight and came crashing down, forming enormous gaping craters called calderas.

Fiery fountains of lava burst through cracks in the thin crust, paving over the Earth with thick layers of basalt, which formed vast basalt plains similar to those on the moon and Mars. The entire surface of the Earth was dotted with numerous active volcanoes erupting one after another in great profusion.

Icy visitors from outer space pounded the infant Earth and supplied substantial quantities of water vapor and gases. The barrage of meteors and comets began around 4.2 billion years ago and continued at a high rate of impact until about 3.9 billion years ago. Other bodies in the solar system, including the Earth's own moon show numerous craters from this massive bombardment (Figure 98). Afterward, the rate of impact remained for the most part about what it is today. This was fortunate, for life would probably have had a poor chance of coming into existence if the Earth was continuously being bombarded by large meteorites.

Some of the meteorites that pounded the Earth

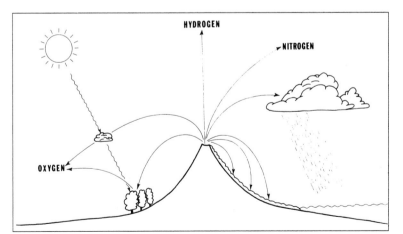

Figure 97 The contribution of volcanoes to the Earth's atmosphere and crust.

Figure 98   A large meteorite crater on the lunar surface. Photo courtesy NASA

were composed of rock, many were metallic, and others were composed of water ice and frozen gas. These along with comets, which are essentially rocky material encased in ice, came from the outer reaches of the solar system. In addition, a great deal of carbon fell out of the sky from primitive meteorites known as carbonaceous chondrites, which are carbon-rich rocks left over from the formation of the Solar System.

With all this heavy volcanic and meteoritic activity, it was not long before the Earth acquired a substantial atmosphere. Water vapor was so heavy that the atmospheric pressure was several times greater than it presently is. The surface was still very hot, and water vapor, carbon dioxide, methane, and ammonia, which broke down into nitrogen and hydrogen, produced a powerful greenhouse effect shrouding the Earth in a thick blanket of steam. Therefore, the early Earth looked very much like Venus does today (Figure 99).

The amount of carbon dioxide in the primordial atmosphere was 1000 times greater than it is today. During the first billion years, the Sun's output was about a third less than it is today. Therefore carbon dioxide acted as a sort of thermal blanket, allowing the Earth to retain its heat. The greenhouse gases kept the temperature of the early atmosphere well above the boiling point of water even though the sun shined only about as much as it does now on Mars. If the Earth had today's atmosphere at that time, it would now be like Antarctica in the dead of winter, and the ocean would be a solid block of ice.

The original oceanic crust was composed of basalt lava flows that erupted on the surface long before the ocean basins began to fill with water. Then around 4 billion years ago, when the Earth finally cooled down, the rains came in torrents, producing the greatest floods the planet has ever known. Deep meteorite craters and volcanic calderas rapidly filled, becoming huge bowls of water that spilled onto flat lava plains. Giant valleys were carved out as water rushed down the steep sides of tall volcanoes, which continued to spew steam and gases into the atmosphere. In addition, multitudes of icy comets continuously pounded the Earth, adding more water to the deluge.

When the rains ended and the skies finally cleared, the Earth was transformed into a glistening blue orb covered almost entirely with a deep global ocean. Vol-

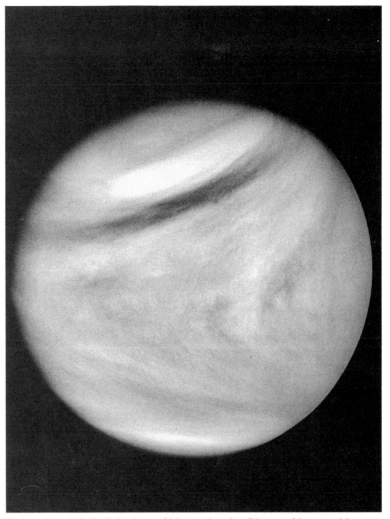

Figure 99   A full-disk view of Venus by the *Pioneer-Venus* orbiter on February 10, 1979, showing the thick cloud layer. Photo courtesy NASA

canoes rising from the ocean floor dotted the seascape with a few scattered islands, but as yet there were no continents. The floor of the ocean was an alien world. Volcanoes continued to erupt undersea, and hydrothermal vents spewed out hot water containing sulfur and other chemicals (Figure 100). In a short time, the sea turned from fresh to salty and contained all the ingredients necessary for the emergence of life. In effect, the ocean became a vast chemical factory, manufacturing all the substances needed to sustain living things.

THE WATER CYCLE

The oceans cover about 70 percent of the Earth's surface with an average depth of about 2.5 miles, amounting to nearly 250 million cubic miles of water. Each day, 1 trillion tons of water rain down on the planet, most of which falls directly into the sea. The movement of water on the planet is one of nature's most important cycles; for without the transport of water over the land and back to the ocean, there would be no life as we know it.

The average journey water takes from the ocean to the atmosphere, across the land, and back to the sea again requires about 10 days. The journey is only a few hours in the tropical coastal regions, but might take as long as 10,000 years in the polar regions. This is what is known as the hydrologic cycle, or simply the water cycle (Figure 101).

The fastest route water can take back to the ocean is by runoff in rivers and streams. This is perhaps the most apparent as well as the most important part of the water cycle. Surface runoff supplies minerals and nutrients to the ocean and cleanses the land. Acidic rainwater reacts chemically with metallic minerals on the surface, producing metallic salts that are carried in solution by rivers emptying into the sea. Rainwater also percolates into the ground, dissolves minerals from porous rocks, and transports these by groundwater.

Solid rock exposed on the surface is broken down chemically into clays and carbonates and mechanically into silts, sands, and gravels. Rivers carry the sediment to the shore. After reaching the ocean, the river's velocity falls off sharply, and the

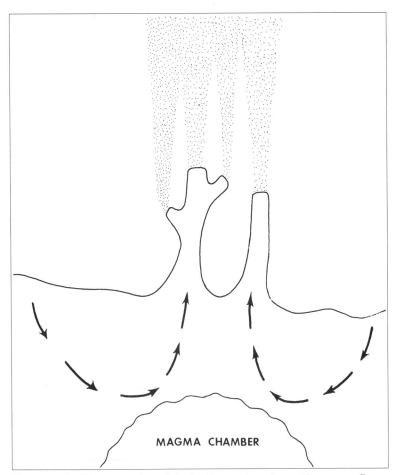

Figure 100   The operation of hydrothermal vents on the ocean floor.

MAGMA CHAMBER

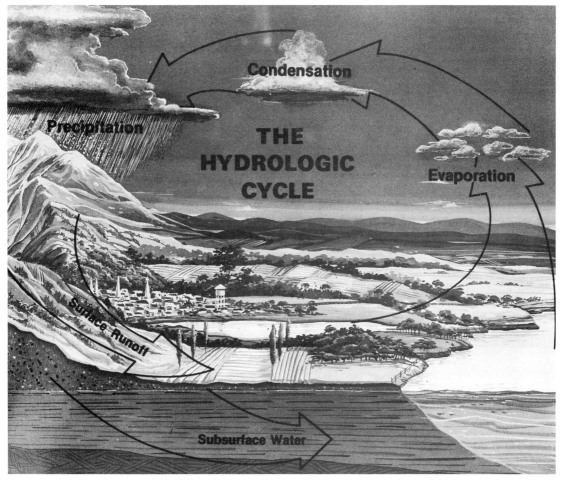

Figure 101   The hydrologic cycle is a continuous flow of water from the ocean over the land and returning to the sea. Illustration courtesy USGS

sediment load drops out of suspension. Chemical solutions carried by the rivers are thoroughly mixed with seawater by currents and wave action. These substances are distributed evenly throughout the ocean with a mixing time of about 1000 years.

The sediments reaching the seashore continually build the continental margins outward. Coarser sediments accumulate near shore, and progressively finer sediments settle out of suspension farther out to sea. As the shoreline advances seaward, the original fine sediments are overlain by coarser sediments. As the shoreline recedes due to rising sea levels, coarse sediments are overlain by fine sediments. This produces a sedimentary sequence of sandstones, siltstones, and shales. In addition, carbonates precipitate and accumulate in thick beds on the shallow ocean floor.

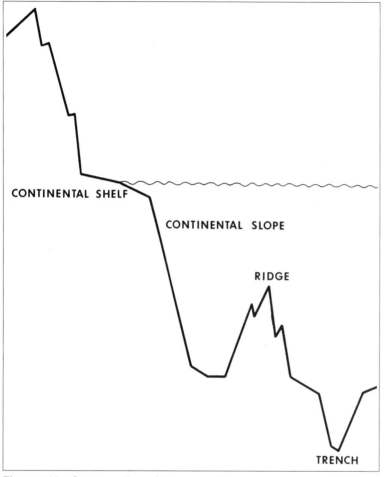

CONTINENTAL SHELF

CONTINENTAL SLOPE

RIDGE

TRENCH

Figure 102   Cross section of the ocean floor.

Most of the sedimentary deposits on the ocean floor are composed of detritus along with shells and skeletons of dead microscopic organisms that flourish in the sunlit waters of the mixed layer of the ocean, the topmost 250 feet. Detritus, which is generated by the weathering of surface rocks along with decaying vegetable matter, is carried by rivers to the edge of the continent and deposited onto the continental shelf (Figure 102), where the material is picked up by marine currents.

When the detritus reaches the edge of the continental shelf, it falls to the base of the continental slope under the pull of gravity. Also, a significant amount of terrestrial material is blown out to sea by dust storms. Approximately 15 billion tons of continental material reach the outlets of rivers and streams annually. Most of the detritus is trapped near the outlets of rivers and on continental shelves, and only a few billion tons actually make it to the deep sea.

Material from dead organisms contributes about 3 billion tons of sediment that accumulate on the ocean floor each year. The rates of accumulation are governed by the rates of biologic productivity, which are controlled in large part by the ocean currents. Nutrient-rich water upwells from the ocean depths to the sunlit zone, where the nutrients are consumed by microorganisms. Areas of high productivity and high rates of accumulation are normally found around major oceanic fronts and along the edges of major ocean currents.

The rate of marine life sedimentation is influenced by the ocean depth. The farther the shells have to descend, the lesser are their chances of reaching the bottom before dissolving in the cold waters of the abyssal. Below the calcium carbonate compensation zone—the depth at which calcium carbonate readily dissolves, about 2 miles deep—much fewer shells become fossils. The preservation of shells as fossils also depends on how quickly the shells are buried by sediments and protected from the corrosive action of seawater.

With the absence of bottom currents and only marine-borne sedimentation, there would be an even blanket of material settling onto the original volcanic floor of the world's oceans. Instead, the rivers of the world also contribute a substantial amount of material that ends up on the deep-ocean floor. The largest rivers of North and South America empty into the Atlantic Ocean, which receives considerably more river-borne sediment than does the Pacific Ocean. The Atlantic is also smaller and shallower than the Pacific, so its marine sediments are buried more rapidly and therefore are more likely to survive than are their Pacific counterparts.

Moreover, the deep-ocean trenches around the Pacific trap much of the material that reaches the western edge, where it is subducted into the mantle. Thus, on average, the floor of the Atlantic contains thicker sediments than the Pacific, amounting to about an inch of accumulation per 2500 years. Furthermore, strong near-bottom currents redistribute sediments in the Atlantic on a larger scale than they do in the Pacific. Undersea storms carry huge quantities of sediment, which is dumped in great heaps on the ocean floor when the storm subsides. This makes the bottom of the ocean one of the most dynamic environments on Earth.

THE CARBON CYCLE

It was not until the development of the theory of plate tectonics with its midocean ridges and subduction zones on the ocean floor that the mystery of the missing carbon dioxide was finally solved. For without some means of removing excess atmospheric carbon dioxide, our planet could become nearly as hot as Venus, whose thick carbon dioxide atmosphere maintains surface temperatures equal to molten lead.

The ocean crust is relatively young, less than 5 percent of the Earth's age. The ocean floor is continuously being created at midocean ridges and destroyed in deep-sea trenches. When the seafloor is forced into the Earth's interior, carbon dioxide is driven out of carbonaceous sediments by the intense heat of the mantle. The molten rock along with its contingent of carbon dioxide works its way upward through the mantle and eventually ends up in the magma chambers of volcanoes and midocean ridges. The eruption of volcanoes and the flow of molten rock from midocean ridges resupplies the atmosphere with new carbon dioxide, making the Earth a great carbon dioxide recycling plant (Figure 103).

Carbon dioxide plays a vital role in regulating the Earth's temperature, and major changes in the carbon cycle profoundly affect the climate. As the early sun gradually became hotter and temperatures on Earth rose higher, more water evaporated from the oceans, which increased rainfall on the land. This speeded up weathering processes, resulting in a loss of atmospheric carbon dioxide, which was converted to limestone on the ocean floor.

The drop in levels of carbon dioxide in the atmosphere kept the Earth from overheating. When temperatures began to fall, less water was evaporated from the ocean, chemical and biologic reactions slowed down, and less carbon dioxide was removed from the atmosphere even though the input from volcanoes and rift zones remained somewhat constant. Thus, the carbon cycle functions as the Earth's thermostat, keeping global temperatures within tolerable limits for life.

The oceans play a major role in regulating the level of atmospheric carbon dioxide. In the upper layers of the ocean, the concentration of gases is in equilibrium with the atmosphere at all times. The upper 250 feet contains as much carbon dioxide as the entire atmosphere. The gas dissolves into the waters of the ocean mainly by the agitation of surface waves. If the ocean were lifeless, much of its reservoir of dissolved carbon dioxide would reenter the atmosphere, more than tripling its present content.

Fortunately, the ocean is teeming with life, and marine organisms take up carbon dioxide in the form of dissolved bicarbonates to build their carbonate skeletons and other supporting structures (Figure 104). When the organisms die, their skeletons sink to the bottom of the ocean, where they dissolve in the deep waters of the abyssal, which holds by far the largest reservoir of carbon dioxide.

In shallow water, the carbonate skeletons build deposits of limestone, dolostone, and chalk, which bury carbon dioxide in the geologic column. The burial of carbonate in this manner is responsible for about 80

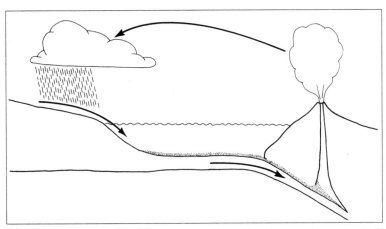

Figure 103   Carbon dioxide converted into bicarbonate is washed off the land by rain and enters the ocean where marine organisms convert it into carbonate sediments, which are thrust into the Earth's interior, and become part of the molten magma. Carbon dioxide returns to the atmosphere by volcanic eruptions.

percent of the carbon deposited on the ocean floor. The rest of the carbonate comes from the burial of dead organic matter washed off the continents. Half of the carbonate is transformed back into carbon dioxide, which eventually escapes into the atmosphere. If it were not for this process, in a mere 10,000 years, all the carbon dioxide would have been taken out of the atmosphere, and the loss of this important greenhouse gas would result in a planet completely covered with ice.

The deep water, which represents about 90 percent of the ocean's volume, circulates very slowly and has a residence time (period of existence) of about 1000 years. It communicates directly with the atmosphere only in the polar regions, so its absorption of carbon dioxide is very limited there. The abyssal receives most of its carbon dioxide in the form of shells of dead organisms and fecal matter that sink to the bottom. The carbon dioxide is returned to the atmosphere by upwelling currents in the tropics. This is why the concentration of atmospheric carbon dioxide is the greatest near the equator.

Because of its large volume, the ocean contains about 60 times more carbon than the atmosphere, mostly in the form of dissolved bicarbonate. Most of the carbon is stored in sediments on the continents and on the ocean floor. The amount of carbon in the form of carbon dioxide in the Earth's original atmosphere might have been as much as one quarter of the total volume. However, as the continents grew, the buildup of sedimentary rock took carbon out of the atmosphere and the

Figure 104 Foraminifera were among the first organisms to build thick deposits of limestone as their shells accumulated on the ocean floor.

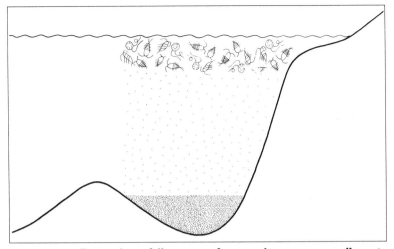

Figure 105   Formation of limestone from carbonaceous sediments deposited on the ocean floor.

ocean and permanently stored it in the crust.

Atmospheric carbon dioxide reacts with rainwater to form a weak carbonic acid, which leaches minerals such as calcium and silica from surface rocks. Rivers transport these minerals to the sea, where they mix with seawater. The minerals are then taken up by marine organisms to make their shells. When the organisms die, their shells sink to the ocean bottom (Figure 105), where they slowly build up deposits of limestone if composed of calcium or diatomite if composed of silica. If the scavenging of carbon dioxide from the atmosphere for the manufacture of carbonaceous sediments on the ocean floor continued unchecked, the atmosphere would soon be depleted of carbon dioxide, and the Earth would be just another cold, lifeless planet.

CARBONATE ROCKS

Carbonate rocks such as limestone, dolostone, and chalk are formed by biologic and chemical precipitation of carbonaceous minerals dissolved in seawater. Carbonic acid is produced by the chemical reaction of water and carbon dioxide in the atmosphere. The acid dissolves calcium and silica minerals from rocks on the surface to form bicarbonates. The bicarbonates enter rivers that empty into the ocean and are thoroughly mixed with seawater by the action of waves and currents.

The bicarbonates precipitate from seawater mostly by biologic activity as well as direct chemical processes. Living organisms use the calcium bicarbonate to build their shells and skeletons composed of calcium carbonate. When the organisms die, their skeletons fall to the bottom of the ocean, where over time the calcium carbonate in the form of a calcite ooze builds up into thick limestone deposits (Figure 106).

The most common precipitate rock is limestone, which is generally produced by biologic processes, as evidenced by an abundance of marine fossils in limestone beds. Some limestone is also chemically precipitated directly from seawater, and a minor amount precipitates in evaporite deposits. Dolomite, which resembles limestone, is produced by the replacement of calcium with magnesium. The mineral is more resistant to acid erosion than limestone, which explains the impressive dolomite peaks in Europe.

Chalk is a soft, porous carbonate rock that erodes easily. Thick chalk beds were laid during the Cretaceous, which is how the period got its name, from the Latin word *creta*, meaning "chalk." The sea has been eroding the thick chalk banks on the Suffolk coast of England for centuries.

Silica also dissolves in seawater in volcanically active areas on the seafloor such as at midocean spreading centers, from volcanic eruptions into the sea, and from weathering of siliceous rocks on the continents. Organisms like diatoms (Figure 107) extract the dissolved silica directly from seawater to make their shells or skeletons. Accumulations of siliceous sediment on the ocean floor from dead organisms form diatomaceous earth, otherwise known as diatomite.

Evaporite deposits are produced in arid regions near shore, where pools of brine, which are replenished with seawater during storms, evaporate and leave salts behind. The salts precipitate out of solution in stages. The first mineral to precipitate is calcite, closely followed by dolomite, although only minor amounts of these minerals are produced in this manner. After about two thirds of the water has evaporated, gypsum precipitates. When nine tenths of the water is removed, halite, or common salt, remains behind.

During its early formation, the ocean became nearly as salty as it is today. Much of the salt removed from the ocean is deposited in thick beds in nearly enclosed basins cut off from the

Figure 106 Intensely folded limestone formation in Atacama Province, Chile. Photo 551 by K. Segerstrom, courtesy USGS

general circulation of the sea. A large portion of the salt is also trapped in seawater between sediment grains on the ocean floor. When thrust deep inside the Earth at subduction zones, the water returns to the surface through volcanic eruptions. Also, seawater percolates through the oceanic crust and picks up minerals along the way. The minerals are then pumped back to the surface by hydrothermal processes, which might help explain why the oceans continue to remain salty.

VOLCANIC ERUPTIONS

Like earthquakes, volcanoes are associated with crustal movements and occur on plate margins. Volcanic eruptions over subduction zones and lava flows at midocean ridges are the final stage of the rock cycle. Volcanoes also play a direct role in regulating the Earth's climate. Large volcanic eruptions spew massive quantities of ash and aerosols into the atmosphere, which blocks out sunlight. Volcanic dust also absorbs solar radiation and indirectly heats the atmosphere, causing thermal disturbances and unusual weather.

There are nearly 400 active volcanoes surrounding the Pacific Ocean that are associated with subduction zones along the rim of the Pacific plate. Tectonic plates are subducted into the mantle by the collision of an oceanic plate with a continental plate or another oceanic plate. The lithosphere melts during its dive into the mantle, and the lighter rock component, due to its greater buoyancy, works its way into the crust to resupply magma chambers with new molten magma.

The composition of the magma also controls the type of eruption. Explosive eruptions occur when a viscous magma containing trapped volatiles and gases is kept from reaching the surface by a plug in the volcano's vent. As pressure increases, the obstruction is

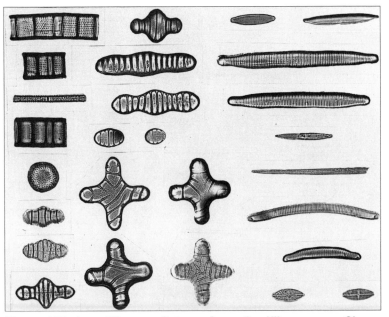

Figure 107 Late Miocene diatoms from the Kilgore area, Cherry County, Nebraska. Photo by G. W. Andrews, courtesy USGS

Figure 108   An active volcano on Andonara Island, Indonesia. Photo courtesy NASA

blown away along with most of the upper peak. Volcanoes associated with subduction zones such as those in the western Pacific and Indonesia (Figure 108) are among the most explosive in the world.

In the Atlantic Ocean, volcanic activity is far less extensive and generally occurs at the Mid-Atlantic Ridge and in the West Indies. Many islands in

Figure 109 **Birth of a new Icelandic island in November 1963, located 7 miles south of Iceland.** Photo courtesy U.S. Navy

the Atlantic are parts of the Mid-Atlantic Ridge system that extends above the sea (Figure 109). Rift volcanoes account for about 15 percent of the world's known active volcanoes, and most are in Iceland and East Africa. Moreover, it is estimated that there are about 20 eruptions of deep submarine rift volcanoes every year. Rift volcanoes on continents such as those in East Africa can be highly explosive.

The output of lava and pyroclastics for a single volcanic eruption varies from a few cubic yards to as much as 5 cubic miles. Rift volcanoes generate about 2.5 billion cubic yards per year of mainly submarine flows of basalt. Subduction zone volcanoes produce about 1 billion cubic yards of pyroclastic volcanic material per year. Volcanoes over hot spots produce about 500 million cubic yards per year, mostly of basalt flows in the oceans and pyroclastics and lava flows on the continents.

Volcanic activity plays an all-important role in restoring the carbon dioxide content of the atmosphere. One of the most important volatiles in magma is carbon dioxide, which helps make it fluid. The carbon dioxide escapes from carbonaceous sediments when they melt after being forced into the mantle at subduction zones near the edges of crustal plates. The molten magma along with its content of carbon dioxide rises to the surface to feed volcanoes that lie on the edges of subduction zones and at midocean ridges. When the volcanoes erupt, carbon dioxide is released from the magma and returned to the atmosphere, and the cycle is complete.

9

TECTONICS AND LIFE

In 1977, while exploring the East Pacific Rise off the tip of Baja California, scientists in the research submarine *Alvin* discovered an oasis 1.5 miles below sea level. Species previously unknown to science were found living in total darkness among hydrothermal vents. Tube worms 3 feet tall swayed back and forth in the hydrothermal currents. Giant crabs scampered blindly across the volcanic terrain. Huge clams up to 1 foot long and clusters of mussels formed large communities around the vents (Figure 111).

The East Pacific Rise is a 6000-mile-long rift system along the eastern Pacific, stretching from north of the Antarctic Circle to the Gulf of California, and is the counterpart of the Mid-Atlantic Ridge. The undersea mountain range lies 2 miles beneath the surface of the sea. At the base of jagged basalt cliffs are evidence of active lava flows, including fields strewn with pillow lava. Exotic-looking chimneys called black smokers spew out hot water blackened with sulfide minerals. Others eject hot water that is milky white and are called white smokers.

The hot water comes from deep below the surface, where seawater percolating down through cracks in the ocean crust comes in contact with magma chambers below spreading centers. It then rises up to the surface and is expelled through hydrothermal vents like undersea geysers (Figure 112). In these volcanically active fields is a world that time forgot. The hydrothermal vents keep the bottom waters at comfortable temperatures, while the surrounding ocean hovers near freezing. The vents also provided valuable nutrients, making this the only environment on Earth that is completely independent of sunlight for its source of energy, which comes instead from the Earth's interior.

The strange creatures discovered near these vents provide only one dramatic example of how species develop in relationship to their environments. Plate tectonics and continental drift have played a prominent role in the history of life almost since the very beginning. Changes in the relative configuration of the continents and the oceans had a far-ranging influence on the environment, climate conditions, and the composition of species. Plate tectonics is therefore among the strongest forces affecting evolutionary changes.

THE BIOSPHERE

Like the Earth, many planets and their satellites possess a core, a mantle, a crust, and even an atmosphere or an icy hydrosphere, but none of them contain a biosphere, which is the living component of the Earth's surface. The biosphere is more than just living things. Life must also be integrated with the geosphere, hydrosphere, and atmosphere to constitute a fully developed biosphere. Biologists have cataloged over 2 million species of plants and animals. Because many species have evaded detection, the total number could easily rise to several times as many, indicating the richness and diversity of life on this planet.

Since life began, it has responded to a variety of chemical climatologic, and geographic changes in the Earth, forcing spe-

Figure 111   Tall tube worms, giant clams, and large crabs live near hydrothermal vents on the deep ocean floor.

Figure 112 Hydrothermal vents on the deep ocean floor provide nourishment and heat for bottom dwellers. Photo courtesy USGS

cies to either adapt or perish. Many dead-end streets along branches of the evolutionary tree are found in the fossil record, which is only a fragmentary representation of all the species that have ever lived. Nearly every conceivable form and function has existed, some more successful than others. It is through this trial-and-error method of specialization that natural selection has chosen certain species to prosper while condemning others to extinction.

Few places on Earth are truly devoid of life. Species are found in the hottest deserts and the coldest polar regions (Figure 113). They reside in the lowest canyons and tallest mountains. They also exist in the deepest oceans and the highest regions of the troposphere. Nor is life excluded from scalding hot springs or deep beneath the ground. Although species most frequently encountered on the Earth's surface seem to play the most dominant role in shaping the planet, it is actually the unseen microscopic creatures, which constitute the largest percentage of the total biomass, that have the greatest influence.

About 80 percent of the Earth's breathable oxygen is generated by photosynthetic single-celled organisms that thrive in the ocean. Microorganisms such as bacteria play a critical role in breaking down the remains of plants and animals for recycling nutrients in the biosphere. Surface plants depend on bacteria in their root systems for nitrogen fixation. Bacteria live symbiotically in the gut of animals and aid in the digestion of food. Biologic processes are responsible for massive concentrations of minerals in the Earth's crust, including silicon, carbon, iron, manganese, copper, and sulfur. Simple organisms also constitute the bottom of the food chain, on which all life ultimately depends for its survival.

Life on Earth constitutes a geologic force that is nonexistent on all other bodies in the Solar System. The evidence of biospheric processes in the Earth's history belongs to the field of biogeology. What appears to be the earliest fossilized remains of microorganisms dates 3.5 billion years ago.

The 3.8-billion-year-old carbonaceous sediments of the Isua Formation in southwest Greenland show a depletion of carbon-13 with respect to carbon-12 determined by the environment, an indication that biologic activity might have existed at a very early age. Therefore, life processes might have been operating for at least four fifths of the Earth's history (Figure 114).

TABLE 11   THE EVOLUTION OF THE BIOSPHERE

Event	Billions of Years Ago	Biologic Consequence	Percent Oxygen	Results
Full oxygen conditions	0.4	Fishes, land plants and animals	100	Approach present biologic conditions
Appearance of shelly animals	0.6	Cambrian fauna	10	Burrowing habitat
Metazoans appear	0.7	Ediacarian fauna	7	First metazoan fossils and tracks
Eukaryotic cells appear	1.4	Larger cells	>1	Red beds, multicellular organisms
Blue-green algae	2.0	Algal filaments	1	Oxygen metabolism
Algal precursors	2.8	Stromatolites	<1	Beginning photosynthesis
Origin of life	4.0	Light carbon	0	Evolution of the biosphere

With this much time involved, it is not surprising that life brought about some dramatic and far-reaching changes to the Earth. The first major alteration was the deposition of banded iron formations on continental margins by iron-metabolizing bacteria. These formations are mined extensively for iron ore around the world. A second major change occurred with the transition of the gas content of the atmosphere and ocean from one-quarter carbon dioxide to one-quarter oxygen by photosynthetic organisms.

Mount Erebus

MC MURDO SOUND

Wright Lower Glacier

Figure 113 Wright Dry Valley, Taylor Glacier region, Victoria Land, Antarctica. Photo 485 by W. B. Hamilton, courtesy USGS

The oxygen in the upper atmosphere promoted the formation of the ozone layer, making conditions safe for plants and animals to conquer the land, which in itself constituted a major change on Earth.

THE GEOSPHERE

The geosphere interacts with the biosphere and directly affects it by changing the environment. There have been 11 episodes of flood basalt volcanism over the past 250 million years. These large eruptions create a series of separate overlapping lava flows that give many exposures world-wide a terracelike appearance, called traps, the Swedish word for stairs. Many flood basalts are located near continental margins, where great rifts separated the present continents from Pangaea. Others, such as the Columbia River basalts of the northwestern United States, are related to hot-spot activity, whereby plumes of hot rocks from deep within the mantle rise to the surface.

The volcanic episodes were relatively short-lived events, with major phases lasting less than 3 million years. Furthermore, the episodes of volcanism appear to be somewhat periodic, occurring about every 32 million years. The timing of these major outbreaks correlates with the occurrence of worldwide mass extinctions of marine organisms. The largest mass extinction in the geologic record occurred at the end of the Paleozoic era 240 million years ago and was responsible for the loss of over 95 percent of all species. It also occurred at about the same time as the eruption of very large traps in Siberia.

During the eruption of a major basaltic lava flow, vigorous fire fountains inject large amounts of sulfur gases into the atmosphere (Figure 115). Moreover, flood basalts release 10 times more sulfur than explosive eruptions. The gases are converted into acid, which has severe climatic and biologic consequences. A major alteration in the composition of the atmosphere can also affect the climate. Volcanoes spew massive quantities of ash and aerosols into the atmosphere, which blocks out sunlight. Volcanic dust also absorbs solar radiation, which heats the atmosphere, causing thermal imbalances and unstable climatic conditions. Heavy clouds of

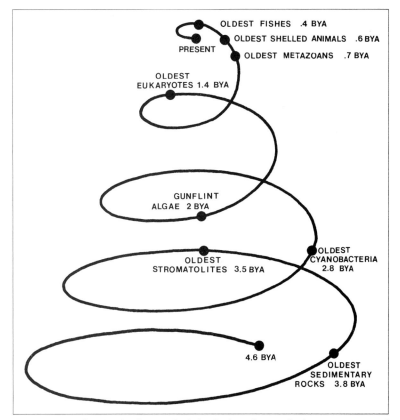

Figure 114 Geologic time spiral depicting the evolutionary stages of life since the Earth began 4.6 billion years ago.

volcanic dust have a high albedo, reflecting much of the solar radiation back into space (Figure 116). This could shade the planet and lower global temperatures. The reduced sunlight might also cause mass extinctions of plants and animals by lowering the rate of photosynthesis.

TABLE 12 FLOOD BASALT VOLCANISM AND MASS EXTINCTIONS

Volcanic Episode	Million Years Ago	Extinction Event	Million Years Ago
Columbian River, USA	17	Low-mid Miocene	14
Ethiopian	35	Upper Eocene	36
Deccan, India	65	Maastrichtian	65
		Cenomanian	91
Rajmahal, India	110	Aptian	110
South-West African	135	Tithonian	137
Antarctica	170	Bajocian	173
South African	190	Pliensbachian	191
Eastern North American	200	Rhaectian/ Norian	211
Siberian	250	Guadalupian	249

A reduction of insolation, or the amount of solar radiation reaching the surface, by 5 percent might result in a drop in global temperatures by as much as 5 degrees Celsius. If maintained long enough, this would be sufficient to initiate an ice age. The long-term cooling would allow glaciers to expand and lower the sea level, which would limit marine habitat area. The lowered temperature could also adversely affect the geographic distribution of species, confining warmth-loving organisms to the tropics.

Extensive volcanic activity 100 times more intense than it is today could produce strong acid rain showers. These in turn could cause widespread destruction of terrestrial and marine species by defoliating plants and altering the acid/alkaline balance of the ocean. Acid gases spewed into the atmosphere might also deplete the ozone layer, allowing deadly solar ultraviolet radiation to bathe the planet, thereby eliminating life on the surface.

Figure 115 Lava fountain and lake at Kilauea Volcano during 1959 and 1960. Photo 58 by D. H. Richter, courtesy USGS

Massive volcanic eruptions might have had a hand in the extinction of dinosaurs (Figure 117). At the end of the Cretaceous period, 65 million years ago, a giant rift opened up on the west side of India and huge volumes of molten lava poured onto the surface. Nearly 500,000 square miles of lava covering an area about the size of France erupted over a period of 500,000 years. The eruptions consisted of about 100 lava flows, which blanketed much of west-central India known as the Deccan Traps (Figure 118) in layers of basalt hundreds of feet thick. It was the largest volcanic catastrophe in the last 300 million years.

An enormous plateau of basalts along the coasts of Greenland and Scotland might have erupted at the same time as the Deccan Traps. There is also evidence of substantial explosive volcanism in an extensive region from the South Atlantic to Antarctica. The eruptions might have dealt a major blow to the climatic and ecologic stability of the planet, forcing many species to become extinct.

When the eruptions began, India was drifting toward southern Asia. The rift separated India from the Seychelles Bank, which was left behind as the subcontinent continued on its journey northward. In the Amirant Basin on the southern edge of the Seychelles Bank about 300 miles northeast of Madagascar lies a remarkably intact circular depression about 200 miles wide, which appears to be a large impact structure. A massive meteorite impact in the Amirante Basin might have triggered the great lava flows that created the Deccan Traps and the Seychelles Islands. Quartz grains

Figure 116   An ash cloud from the 1980 eruption of Mount St. Helens. Photo courtesy NOAA

shocked, or fragmented, by the high pressures generated by the impact were found lying just beneath the immense lava flows, suggesting that they might be linked to the impact. It is possible that such large meteorite impacts create so much disturbance in the Earth's thin outer crust that they can induce massive volcanic eruptions.

Shocked quartz and iridium, an isotope of platinum, found in the Cretaceous-Tertiary (K-T) boundary clay through-

Figure 117   All species of dinosaurs became extinct due to a number of suspected causes, including terrestrial as well as extraterrestrial influences.

out the world (Figure 119) are though to be fallout from a giant meteorite impact. They might also have resulted from massive volcanic eruptions taking place over a period of hundreds of thousands of years. Explosive volcanic eruptions such as the 1980 eruption of Mount St. Helens in southwestern Washington can produce similar shocked quartz grains. Volcanoes whose magma source lies deep within the mantle such as Kilauea on Hawaii can produce significant quantities of iridium, an isotope of platinum that is rare in the crust. The Deccan eruptions could have emitted large amounts of iridium, which would account for the anomalously high concentrations in the K-T boundary layer. Moreover, the microspherules (small, glassy beads) found at the K-T contact that are believed to be the product of impact melt could also have been produced by large volcanic eruptions.

TECTONICS AND EVOLUTION

The motions of the continents had a major influence on the distribution, isolation, and evolution of species. The changes in continental configuration greatly affected global temperatures, ocean currents, productivity, and many other factors of fundamental importance to life. Many different environments result in a wide variety of species. Therefore, evolutionary trends varied throughout geologic time in response to major environmental changes as natural selection acted to adapt organisms to the new conditions

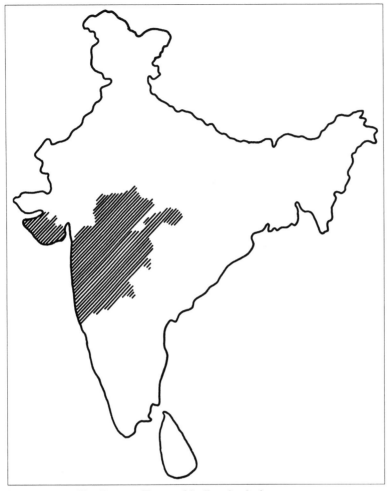

Figure 118   The Deccan Traps of India, shaded.

forced on them by environmental factors that were affected by, among other things, continental drift.

When all the continents were welded into the supercontinent Pangaea near the end of the Paleozoic era (250 mya), there was a great diversity of plant and animal life in the ocean as well as on land (Figure 120). The formation of Pangaea marked a major turning point in evolution of life, during which the reptiles emerged as the dominant species. It is believed that most of the Pangaen climate was equable with few temperature extremes and fairly warm throughout the year. Much of the interior of Pangaea, however, was a desert, whose temperatures fluctuated wildly between seasons, with scorching summers and freezing winters. This temperature fluctuation might have contributed to the widespread extinction of land-based species during the late Paleozoic. It also explains why the reptiles, which adapt readily to hot, dry climates, replaced the amphibians as the dominant land species.

The vast majority of marine species live on continental shelves or shallow-water portions of islands and subsurface rises at shallow depths generally less than 600 feet. The richest shallow-water faunas live in the tropics, which contain large numbers of highly specialized species. Progressing to higher latitudes, diversity gradually falls off, until in the polar regions there are less than one tenth as many species than in the tropics. Moreover, there is twice the species diversity in the Arctic Sea, which is surrounded by continents, than in the Antarctic Sea, which surrounds a continent.

TECTONICS AND LIFE

Diversity is greatly affected by seasonal changes such as variations in surface and upwelling ocean currents that affect the nutrient supply. This in turn causes large fluctuations in productivity. Therefore, the greatest diversity among species is off the shores of small islands or small continents facing large oceans, where fluctuations in the nutrient supply are least affected by seasonal effects of landmasses which often sponsor episodes of glaciation or other environmental swings.

Species diversity also depends on the shapes of the continents, the width of shallow continental margins, the extent of inland seas, and the presence of coastal mountains, all of which are affected by continental motions. When the continents were assembled into Pangaea, there was a continuous shallow-water margin running around the entire perimeter of the supercontinent, with no major physical barriers to the dispersal of marine life.

Figure 119   Geologists point out the Cretaceous-Tertiary boundary at Brownie Butte outcrop, Garfield County, Montana. Photo by B. F. Bohor, courtesy USGS

Furthermore, the seas were largely confined to the ocean basins, leaving the continental shelves mostly exposed. Consequently, habitat area for shallow-water marine organisms was very limited, which accounted for the low species diversity. As a result, marine biotas were more widespread but contained comparatively fewer species.

Similar conditions might have occurred during the late Precambrian around 600 million years ago, when another supercontinent was in existence. During the Cambrian period, it broke into perhaps four or five large continents, whose extended shoreline might have effected the explosion of new species during that time. There were twice as many phyla (organisms that share the same general body plan) during the Cambrian than before or since. Never were there so many experimental, or unusual, organisms (Figure 121), none of which have any counterparts in today's living world. When Pangaea broke up and the resulting continents migrated to their current positions, diversity again increased to unprecedented heights, providing our present-day world with a rich variety of species. The movement of continents changes the shapes of ocean basins, which affects the flow of ocean currents, the width of continental margins, and the abundance of marine habitats. When a supercontinent breaks up, more continental margins are created, the land lowers, and the ocean rises,

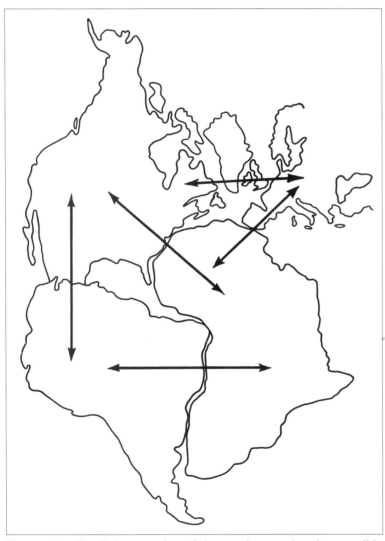

Figure 120   The fitting together of the continents, showing possible migration routes of plants and animals.

providing a larger habitat area for marine organisms.

During times of highly active continental movements, volcanic activity increases. This is especially true at midocean spreading centers, where crustal plates are pulled apart by upwelling magma from the upper mantle, and at subduction zones, where crustal plates are forced into the Earth's interior and remelted to provide the raw materials for new crust. The amount of volcanism could affect the composition of the atmosphere, the rate of mountain building, and the climate, all of which invariably affects life.

The drifting of the continents during the Cenozoic isolated many groups of mammals and they evolved along independent lines. Around 40 million years ago, Australia drifted away from Antarctica, which acted as a bridge between it and South America, and the separation isolated the continent from the rest of the world. It is now inhabited by strange egg-laying mammals called monotremes, including the spiny anteater and platypus, which should rightfully be classified as surviving mammal-like reptiles. When the platypus was first discovered, it was thought to be a missing link between mammals and their ancient ancestors.

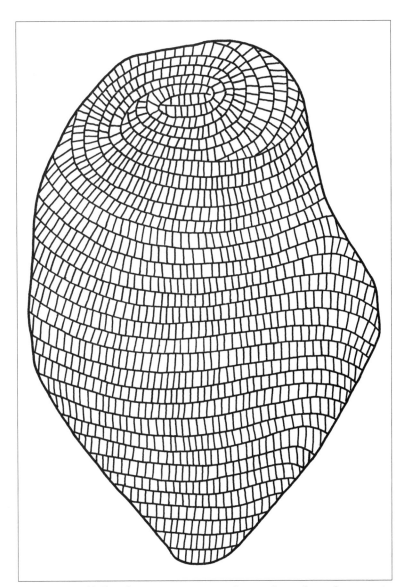

Figure 121 Helicoplacus was an experimental species whose parts were assembled in a way not found in any living creature and became extinct about 510 million years ago.

Marsupials are believed to have originated in North America around 100 million years ago. They migrated to South America, crossed over to Antarctica when the two continents were still attached to each other, and landed in Australia before it broke away from Antarctica. The Australian group consists of kangaroos, wombats, and bandicoots, while opossums and related species occupy other parts of the world.

Camels originated in North America about 25 million years ago and migrated out of the continent to other parts of the world by crossing over connecting land bridges (Figure 122). Horses, which evolved at about the same time, took a similar route out of North America. Madagascar, which broke away from Africa about 125 million years ago, has none of the large mammals that occupy the mainland except for the hippopotamus, which somehow managed to cross over to the island after it had already drifted some distance away from the continent.

TECTONICS AND EXTINCTION

Considering all the great upheavals in the Earth throughout geologic history, it is a great wonder how life managed to survive to the present. Episodes of mass extinctions of species correlate with cycles of terrestrial phenomena. The most pervasive of these is a 300-million-year cycle of convection currents in the Earth's mantle. Convection within the mantle is the very driving force behind plate tectonics and all geologic activity taking place on the Earth's surface.

During the periods of rapid mantle convection, supercontinents tend to break up. This leads to the compression of ocean basins, causing a rise in sea level and a transgression of the seas onto the land. It also increases volcanism, which in turn increases the carbon dioxide content of the atmosphere, resulting in a strong greenhouse effect and rising global temperatures.

During times of low mantle convection, all the continents are assembled into a supercontinent. This causes a widening of ocean basins with a consequent drop

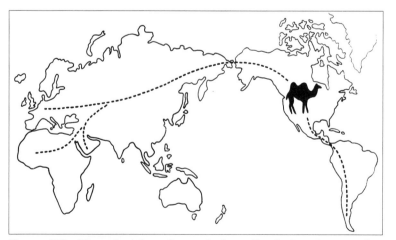

Figure 122 The migration of camels from North America to other parts of the world.

in global sea levels and a regression of the seas from the land. The fluctuation in sea levels is directly related to vertical tectonic movements on the continents. As the continents rise, the ocean lowers. Moreover, there is a reduction of atmospheric carbon dioxide due to low levels of volcanism, resulting in lower global temperatures.

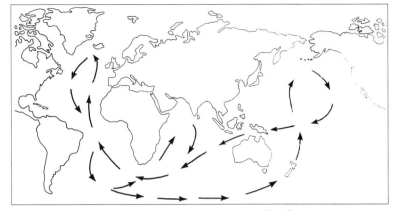

Figure 123   The ocean heat transport system distributes warm water from the tropics to cold regions of the world.

Throughout most of the Earth's history, continents and ocean basins have been continuously reshaped and rearranged by crustal plates in motion. When continents break up, they override ocean basins, which makes the seas less confined and raises global sea levels several hundred feet. Low-lying areas inland of the continents are inundated by the sea, dramatically increasing the shoreline and the shallow-water marine habitat area, which can support a larger number of species.

There is also a great deal of mountain building associated with the movement of crustal plates. This alters patterns of river drainages and climate, which in turn affects terrestrial habitats. Raising land to higher elevations, where the air is thinner and colder, allows glaciers to form, especially in the higher latitudes. Continents scattered in all parts of the world can also interfere with ocean currents, which affect global heat distribution (Figure 123).

When continents are assembled into a supercontinent, land no longer impedes the flow of ocean currents. They can therefore distribute heat more evenly around the planet and maintain more uniform global temperatures. The ocean basins also widen, causing sea levels to drop. A substantial and rapid fall in sea level could directly and indirectly influence the biologic world. It would increase the seasonal extremes of temperature on the continents, thereby increasing environmental stress on terrestrial species.

A drop in sea level also forces inland seas to retreat, resulting in a continuous, narrow continental margin around the supercontinent. This in turn reduces the shoreline, which radically limits the habitat area. Moreover, unstable near-shore conditions result in an unreliable food supply. Many species cannot cope with the limited living space and food supply and die out in tragically large numbers. This occurred at the end of the Paleozoic era, when the vast majority of marine species became extinct.

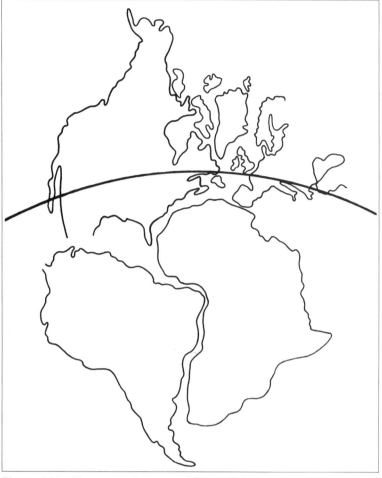

Figure 124   The approximate positions of the continents relative to the equator during the carboniferous period.

During the final stages of the Cretaceous period, when the seas were departing from the land and the level of the oceans began to drop, the temperatures in a broad tropical ocean belt known as the Tethys Sea began to fall. This might explain why the Tethyan species that were the most temperature sensitive suffered the heaviest extinction at the end of the period. Species that were amazingly successful in the warm waters of the Tethys were totally decimated when temperatures dropped. Afterward, marine species took on a more modern appearance as ocean bottom temperatures continued to plummet.

TECTONICS AND CLIMATE

The positions of the continents also determined global climatic conditions. Over the last 100 million years, changes in climate have been largely controlled by plate tectonics. When most of the land huddled near the equatorial regions, (Figure 124) the climate was warm. When lands wandered into the polar regions, however, they grew cold and became covered with ice. Furthermore, during times of highly active continental movements, there is a greater amount of volcanic activity, especially at midocean ridges and subduction zones. The amount of volcanism could affect the composition of the atmosphere and the rate of mountain building, which ultimately affect the climate.

When all the lands were welded into the supercontinent Pangaea around 250 million years ago, much of the land was located near the tropics, where more of the sun's heat could be absorbed, which contributed to higher global temperatures. Also, oceans in the high latitudes are less reflective than land. Therefore, they could absorb more heat, which further moderated the climate. Moreover, without land in the polar regions to interfere with the movement of warm ocean currents, both poles remained ice free year-round. This resulted in little variation in temperature between the high latitudes and the tropics.

When Pangaea began to break up, the climate in the Cretaceous period was extremely warm, with average global temperatures 5 to 10 degrees Celsius warmer than they are today. When the continents drifted toward the poles by the end of the Cretaceous period, however, they disrupted the transport of poleward oceanic heat and replaced heat-retaining water with heat-losing land. As the cooling progressed, the land accumulated snow and ice, creating a greater reflective surface, which further lowered global temperatures and sea levels. The most important factor controlling the geographic distribution of marine species is ocean temperature. Therefore, climatic cooling is primarily responsible for most crises among seafaring creatures.

The location of land in the polar regions is often the cause of extended periods of glaciation because high-latitude land has a higher albedo and a lower heat capacity than the surrounding seas. This condition encour-

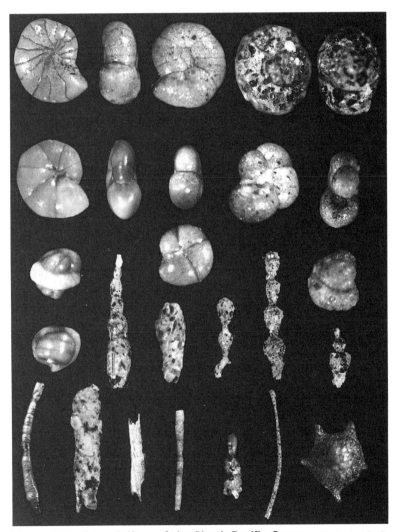

Figure 125 Foraminifera of the North Pacific Ocean. Photo by P. B. Smith, courtesy USGS

ages the accumulation of snow and ice. The more land area in the higher latitudes, the colder and more persistent is the ice, especially when much of the land is at higher elevations, where glaciers can grow easily. Furthermore, replacing land in the tropics with oceans has a net cooling effect because land located in the lower latitudes absorbs more of the sun's heat, while oceans reflect solar rays back into space. Also, by increasing the land area in the high latitudes, where snow falls steadily with little melting, a permanent polar glacial climate is established.

After the glaciers are securely in place, the high reflectivity of snow and ice tends to perpetuate them and sustain glaciation. This is true even if the once high land were to sink to the level of the sea due to a decrease in crustal buoyancy by the weight of the ice. Increasing the weight on the land can also interfere with the flow of mantle currents, possibly causing an increase in volcanism. The greater volcanic activity increases the amount of ash injected into the atmosphere, which shades the planet and lowers global temperatures, maintaining glacial conditions.

The ocean bottom influences how much heat is carried by ocean currents from the tropics to the poles. When Antarctica separated from South America and Australia and moved over the South Pole some 40 million years ago, a circumpolar Antarctic ocean current was established. This current isolated the frozen continent, preventing it from receiving warm poleward flowing waters from the tropics, causing it to become a frozen wasteland. During this time, warm salty water filled the ocean depths, while cooler water covered the upper layers.

In the large shallow Tethys Sea, which separated Eurasia from Africa, warm water became top-heavy with salt due to high evaporation rates and little rainfall, causing it to sink to the bottom. Meanwhile, ancient Antarctica, which was much warmer than it is today, generated cool water that filled the upper layers of the ocean, causing the entire ocean circulation system to run backward. Then around 28 million years ago, Africa slammed into Eurasia turning off the tap of warm water flowing to the poles, placing Antarctica in a deep freeze. The cold air and ice cooled the surface waters and made them heavy enough to sink and flow toward the equator, providing the Earth with the ocean circulation system it has today.

Changes in the deep-ocean circulation also coincided with the late Eocene extinctions, in which all the archaic mammals abruptly disappeared. The extinctions also eliminated many European species of marine life, when the continent was flooded by shallow seas. The separation of Greenland from Europe during this time might have been responsible for the frigid arctic waters to drain into the atmosphere, significantly lowering its temperature, and causing most types of foraminifera to disappear (Figure 125). In the ensuing Oligocene epoch, the seas were drained from the land as the ocean withdrew to perhaps its lowest level since the last several hundred million years, and the seas remained depressed for the next 5 million years.

These cooling events removed the most vulnerable of species, so that those living today are more robust and capable of withstanding the extreme environmental swings that occurred during the last 2 million years, when glaciers spanned across much of the Northern Hemisphere (Figure 126). The ice ages came

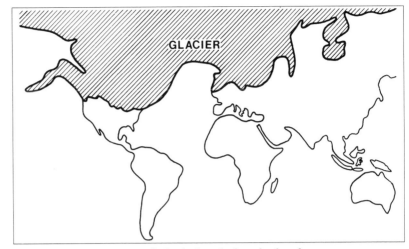

Figure 126   The extent of glaciation during the last ice age.

and went almost like clockwork about every 100,000 years. They will most likely continue for perhaps several more million years until all the continents are once again gathered at the equator. This will alter the face of the planet and the composition of its life, just as it has done for most of the Earth's history.

10

TECTONICS IN SPACE

The geologic activity on Earth, driven by heat flowing from the interior, constantly reshapes the surface as crustal plates collide, volcanoes expel hot rock and gases through cracks in the crust, and as earthquakes wrench the Earth's crust at plate boundaries. The Earth is not the only body in the Solar System to possess this activity; the inner terrestrial planets, Mercury, Venus, and Mars, had similar origins. Except for the Earth and possibly Venus they are now tectonically dead. Their volcanoes have long since ceased erupting, and there are no longer any quakes. But several moons of the outer planets appear to remain active. Volcanic eruptions have actually been observed on Jupiter's moon Io. Indeed, the moon is so active it rivals even the Earth in its fusillade of fiery blasts.

LUNAR TECTONICS

Our Earth and its moon appear to have formed roughly about the same time some 4.6 billion years ago. The moon has a crust 30 to 60 miles thick, making it over twice as thick as the crust of the Earth, and a brittle mantle about 700 miles thick. Below that is a relatively dense core with a radius of about 300 miles that might be partly molten.

Early in its history, the moon surface was melted by a massive meteorite bombardment, which was responsible for most of its present-day terrain features (Figure 127). Numerous large impacts producing craters up to 250 miles across destroyed most of the moon's original crustal rocks. Beginning about 4.2 billion years ago, and continuing for a period of several hundred million years, huge basaltic lava flows welled up through the weakened crater floors and flooded great stretches of the moon's surface, filling and submerging many of the meteorite craters.

The dark basalts covered some 17 percent of the lunar surface lava that hardened into smooth plains called maria, meaning seas, named so because when viewed from Earth they look like seas. The composition of the basalts indicates that they came from a deep-seated source, much deeper than magmas from the Earth's interior. This resulted in massive outpourings of molten rock onto the surface, producing oceans of basaltic lava.

There are numerous narrow, sinuous depressions, or rills, in the lava flows. Many of these emanate from craters that appear to be volcanic in origin. In some areas, wrinkles break the surface of the lava flows, which might have been caused by moonquakes, which explains the extensive seismic activity sensed by lunar probes. There are also areas of previous volcanic mountain building throughout the moon's surface, with ridges reaching several hundred feet high and extending for hundreds of miles. The last of the basalt lava flows hardened about 3 billion years ago, and except for several fresh meteorite impacts the moon looks much the same today as it did then.

Moon rocks bought back during the Apollo missions (Figure 128) vary in age from 3.2 to 4.5 billion years old. The oldest rocks are primitive and have not

Figure 127   A view of a full moon taken from the *Apollo 11* spacecraft, showing numerous craters. Photo courtesy NASA

Figure 128 Moon rock brought back by Apollo 15. Photo courtesy NASA

changed significantly since they first originated from molten magma. These rocks formed the original lunar crust and for this reason are called Genesis Rock. They are composed of a coarse-grained, feldspar granite formed deep in the moon's interior. The youngest rocks are volcanic in origin and were melted and reconstructed by giant meteorite impacts. When the basalt flows ended, new moon rock formation appears to have ceased, and no rocks are known to be younger than 3.2 billion years.

All lunar rocks are igneous, meaning they were derived from molten magma, and together they form the regolith, which is composed of loose rock material on the surface. The rocks include coarse-grained gabbro basalt, meteoritic impact breccia, pyroxine peridotite, glass beads known as chondrules derived from meteorite impacts, and dust-size soil material. The regolith is generally about 10 feet thick but is thought to be thicker in the highlands. Because of its dark basalts, the moon's surface is a poor reflector of sunlight. Only about 7 percent of the light is reflected, which makes the moon one of the darkest bodies in the Solar System.

MERCURIAN TECTONICS

If Mercury and the moon were placed side by side, there would be many striking resemblances. Photographs of Mercury taken during the *Mariner 10* encounter in March 1974 (Figure 129) could easily be mistaken for the back side of the moon. About the only major difference is that Mercury does not have jumbled mountainous regions and wide lava plains, or maria. Instead, it has long, low, winding cliffs that resemble fault lines several hundred miles long. Like the moon, Mercury is heavily scarred with meteorite craters from a massive bombardment between 4.2 and 3.9 billion years ago, and little has happened since.

The compositions of Mercury and the moon are very similar as well and comparable to the interior of the Earth. Mercury has an appreciable magnetic field, indicating the presence of a large metallic core, which also accounts for the planet's high density. Above the core lies a silicate mantle, constituting only a quarter of the radius, whereas the Earth's mantle takes up nearly half the radius and nearly 70 percent of the planet's volume.

Figure 129 A simulated encounter with Mercury by *Mariner 10* in March 1974. Photo courtesy NASA

Mercury has the widest temperature extremes of any planet. During the day, temperatures soar to 300 degrees Celsius, hot enough to melt lead. At night, temperatures plummet to -150 degrees Celcius. This is mainly due to the planet's closeness to the sun, a highly elliptical orbit that ranges between 29 and 43 million miles from the sun, a slow rotation rate of 1.5 times for every revolution around the sun (which keeps its dark side away from the sun for long periods), and a lack of an appreciable atmosphere to spread the heat around the planet. Mercury revolves around the sun every 88 Earth days and completes one rotation on its axis every 59 Earth days. This unusual orbital feature requires Mercury to go twice around the sun before completing a single day.

Any gases and water vapor formed by volcanic outgassing or degassing from meteorite impacts were quickly boiled off due to the excessive heat and low escape velocity needed to break away from the low gravity. Because of its small size, which allowed all the planet's internal heat to escape into space early in its life, Mercury is now a tectonically dead planet.

VENUSIAN TECTONICS

Venus is the Earth's sister planet in many ways. It is nearly the same size and mass and contains a substantial atmosphere composed almost entirely of carbon dioxide. As on Earth, the geology on Venus still appears to be highly active. Surface mapping of Venus by Earth-based radar and radar observations from spacecraft in orbit around the planet have revealed a crumpled and torn surface that is typical of jostling segments of crust comprising movable tectonic plates. The horizontal motion of the crust has compressed it into ridges and troughs comparable only to those found on Earth.

Folded and broken crust, formed by the collision of tectonic plates, built mountains. Some of these bear a considerable resemblance to the Appalachians, which were formed by the collision of the North American and African continents. The 36,000-foot-high Maxwell Montes dwarfs Mount Everest by over a mile. Faults resembling San Andreas shoot through the surface, displacing large chunks of crust. Although Venus possesses continental highlands in the northern hemisphere, its ocean basins lack one essential ingredient—water.

Venus also appears to be very volcanically active, much more so than Earth, and might resemble our planet during its early stages of development. The surface of Venus is relatively young, with an average age of less than 1.5 billion years, and is comparable to the Earth of 3 billion years ago before plate tectonics began to operate extensively. Radar maps made by data supplied by orbiters along with surface elevations from Earth-based radar have revealed large volcanic structures on the surface of Venus. Very low frequency radio waves called "whistlers" picked up by the orbiters are believed to have originated when lightning arcs across volcanic dust clouds.

A region known as Beta Regio appears to possess numerous large volcanoes, some of which are up to 3 miles high. A large shield volcano known as Theia Mons has a diameter of over 400 miles, much larger than any volcano on Earth. Large anomalies in the gravity field seem to indicate that some highlands on Venus are buoyed up by rising plumes of mantle rock, which are thought to feed active volcanoes. An abundance of sulfur gases in the atmosphere suggests that Venus has recently undergone massive volcanic eruptions.

Radar mapping has revealed large elevated plateaulike tracts, rising 3 to 6 miles above the surrounding terrain. There are elongated ridges and circular depressions over 50 miles in diameter that might have resulted from large meteorite impacts. A great rift valley measuring 8 miles deep, 175 miles wide, and 900 miles long might well be the grandest canyon in the Solar System (Figure 130). The jagged surface on Venus appears to have been shaped by deep-seated tectonic forces and volcanic activity, producing a landscape that is totally alien to anything found on Earth.

The surface on Venus appears to be remarkably flat (Figure 131), much more so than Earth or Mars. Two thirds of the planet has a relief of less than 3000 feet. Scattered rocks on the surface observed by Soviet landing craft were angular in some places, but flat and rounded at other locations, which suggests strong wind erosion. The surface rocks have a density identical to terrestrial granites. The soil has a composition similar to basalts on Earth and our moon. One curious aspect about Venus, however, is that it has no moon, even though the planet's origins are thought to be very similar to the Earth. This might indicate that the creation of our moon was a unique event in the evolution of the Solar System.

The internal structure of Venus is thought to be much like the inner workings of the Earth. Venus possesses a liquid metallic core surrounded by a mantle and a rocky crust. The core constitutes about one quarter of the planet's mass and half its radius. There is almost a complete absence of a magnetic field, however. This might be due to the planet's slow rotation and lack of an

Figure 130   An artist's impression of the Venus rift valley, which is 3 miles deep, 175 miles wide, and 900 miles long, the largest canyon yet found in the Solar System. Illustration courtesy NASA

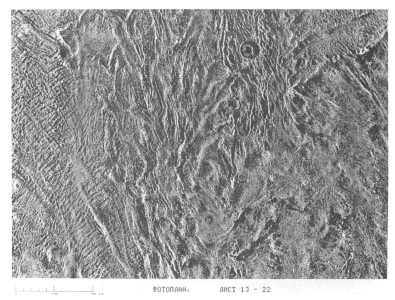

Figure 131   Radar images of Venus from *Vernera* spacecraft. Photo courtesy NASA

inner core, which prevents the generation of the electric currents needed to produce a magnetic field.

The density of the Venusian mantle ranges from 5.6 times the density of water near the top to 9.5 times near the core. The density of the crust averages 2.9, similar to the Earth's crust. The presence of the radioactive elements uranium, thorium, and potassium in the crust are in amounts comparable with those found on Earth. Therefore, it is reasonable to assume that the amount of radiogenic heat in the interior of Venus is comparable to that of the Earth's interior. As with the Earth, the best escape for excess internal heat is volcanism.

Upwelling of magma below the volcanic region might be responsible for horizontal movements in the crust in much the same manner that global tectonics operates on Earth. This has created a major linear rift system that is flanked by volcanic structures. Isolated peaks have also been identified near the equator, and these structures are thought to be individual volcanoes.

Orbiting spacecraft have observed large circular features as wide as several hundred miles and relatively low in elevation. These structures are attributed to gigantic volcanic domes that have collapsed, leaving huge, gaping holes called calderas surrounded by folds of crust as though massive bubbles of magma had burst on the surface.

MARTIAN TECTONICS

Like the moon and Mercury, the surface on Mars is highly cratered. However, Mars also has terrain features that have been produced by wind, water, and ice erosion as well as tectonic activity, including massive volcanic eruptions. Mars is only about half the size of the Earth, and like the Earth's moon it has been tectonically dead for what appears to be well over 2 billion years. However, the darkness of some pyroclastics suggests that they are no older than a few million years. Therefore, Mars could have been a volcanically active planet throughout most of its history.

Branching tributaries running across the Martian surface look similar to dry river beds. Apparently, heat generated either by volcanic activity or by meteorite bombardments melted subterranean ice, resulting in great floods of water and flowing mud gushing forth and carving out gigantic ditches that rival anything found on Earth. The largest canyon, Valles Marineris (Figure 132), measures 3000 miles long, 100 miles wide, and 4 miles deep and could hold several Grand Canyons with room to spare. The canyon is thought to have formed by slippage of the crust along giant faults similar to those of the East African Rift Valley. This activity is thought to have been accompanied by volcanism just like that on Earth.

Images returned by orbiting spacecraft revealed a Martian northern hemisphere with numerous volcanoes. The largest of these is Olympus Mons (Figure 133), which covers an area about the size of Ohio. It has an elevation of

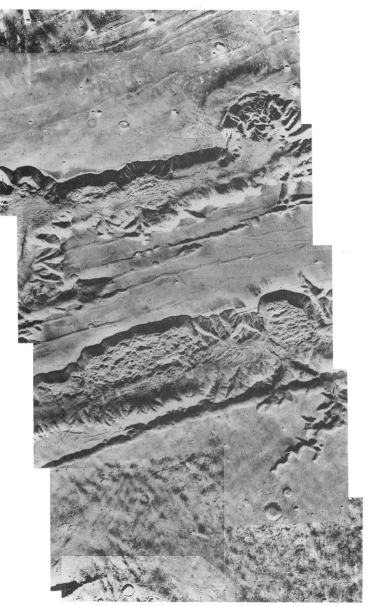

Figure 132 A mosaic of the Martian surface at the west end of the Valles Marineris canyon system. These two canyons are over 30 miles wide and nearly 1 mile deep. Photo courtesy NASA

75,000 feet, over twice as high as Mauna Kea, Hawaii, the tallest volcano on Earth. The Martian volcanoes closely resemble shield volcanoes similar to those that built the main island of Hawaii. Their extreme size might have resulted from the absence of plate movements. Rather than forming a chain of relatively small volcanoes as though they were assembled on a conveyer belt when a plate moved across a volcanic hot spot, a single very large cone developed due to the crust remaining stationary over a magma body for long periods.

Impact craters are notably less abundant in the regions where volcanoes are most numerous, indicating that much of the volcanic topography on Mars formed after the great meteorite bombardment some 4 billion years

Figure 133   A mosaic of the 350-mile-diameter Olympus Mons and surrounding Martian countryside.
Photo 439 by E. C. Morris, courtesy USGS

ago. The southern hemisphere has a surface that is more highly cratered and is comparable in age to the lunar highlands, which are about 3.5 to 4.0 billion years old. The discovery of several highly cratered and weathered volcanoes indicates that volcanic activity began early and had a long history. However, even the fresh-appearing volcanoes and lava planes of the northern hemisphere are probably very ancient.

TABLE 13 MAJOR VOLCANOES OF MARS

Volcano	Height (miles)	Width (miles)	Age (million years)
Olympus Mons	16	300	200
Ascraeus Mons	12	250	400
Pavonis Mons	12	250	400
Arsia Mons	12	250	800
Elysium Mons	9	150	1000–2000
Hecates Tholus	4.5	125	1000–2000
Alba Patera	4	1000	1000–2000
Apollinaris Patera	2.5	125	2000–3500
Hadriaca Patera	1	400	3500–4000
Amphitrites Patera	1	400	3500–4000

Seismographs on Mars landers detected no seismic activity, indicating that the planet might be tectonically dead. There appears to be no horizontal crustal movement of individual lithospheric plates, and the crust is probably much colder and more rigid than the Earth's crust. This explains why there are no folded mountain ranges on Mars. Because it has far less mass than the Earth, Mars could not have generated and stored enough heat to keep its interior in a semimolten state. Furthermore, Mars has no appreciable magnetic field because it either has no metallic core or the core is not fluid due to a low internal temperature.

Because there is no evidence of plate tectonics, heat could not be generated by friction between plates. Therefore, in order for Mars to exhibit recent volcanism under these conditions, pockets of heat would have to exist just below the surface. Evidence for polar wandering on Mars indicates that the entire crust might have shifted as a single plate due to instabilities in the interior resulting from the planet's rotation. This movement could have been caused by convection currents in the mantle, forcing molten rock to the surface, where great outpourings of magma produced maria similar to the vast lava plains on the moon.

JOVIAN TECTONICS

The large outer bodies of the Solar System are referred to as the giant gaseous planets, and because of their low densities it is believed that gases make up the bulk of their masses. This alone indicates that these bodies are completely different from the inner, dense, terrestrial planets. Although the large planets are spectacular in their own right, they are not as geologically impressive as their moons. *Voyager 1* and *Voyager 2* spacecraft passed by Jupiter in January and June 1979, respectively, and obtained fantastic images of the planet and its larger inner moons (Figure 134).

Jupiter and its satellites resemble a miniature solar system comprising 15 moons. The four largest moons, first discovered by the Italian scientist Galileo in 1610, travel in nearly circular orbits with periods ranging from 2 to 17 days. The largest of the Galilean moons, Callisto and Ganymede, are about the size of Mercury. The two smaller moons, Europa and Io, are about the size of the Earth's moon. The surface of Callisto, the outermost of the Galilean satellites, is densely cratered like the back side of

Figure 134   Jupiter and its four planet-size moons as viewed from *Voyager 1* in March 1979. Photo courtesy NASA

the Earth's moon, except the meteorite impacts were made on a frozen crust of dirty ice.

Ganymede is the largest moon in the Solar System. Its surface resembles the near side of our moon with densely cratered regions and smooth areas, where young lava flows covered the scars of older craters. Numerous grooves suggest that some type of tectonic activity has taken place in the not too distant past (Figure 135).

Europa has a smooth icy surface, devoid of major impact craters and is crisscrossed by numerous linear features that are thousands of miles long and 100 miles wide. These might be fractures in the icy crust that were filled with material erupted from below. The moon probably formed after the great meteorite bombardment period, 4 billion years ago, when impacts were much more prevalent than they are today.

The most intriguing of the Jovian moons is Io (pronounced EYE-oh). It is the innermost moon, and its size, mass, and density are nearly identical to the Earth's moon. Due to a gravitational tug of war between Jupiter and Ganymede, with Io caught in the middle, frictional heat has kept its interior in a constant molten state with temperatures above 370 degrees Celcius. This produces widespread volcanism over the moon's entire surface, which has over 100 active volcanoes (Figure

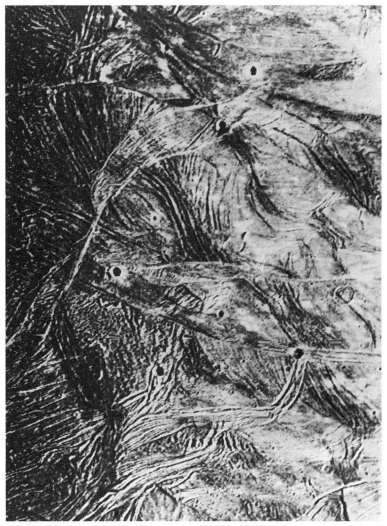

Figure 135 Ganymede is Jupiter's largest moon and shows ridges and grooves that probably resulted in deformation of the icy crust.
Photo courtesy NASA

136). Nine volcanoes were actually erupting when *Voyager 1* flew by in early 1979. This makes Io possibly the most volcanically active body in the Solar System. Io's volcanism also produces a surface that is highly colorful due to the multicolored sulfurous rocks, which paint the moon in hues of yellow, orange, and red.

The almost total lack of impact craters, indicates that the surface of Io has recently, within the last million years, been paved over with large amounts of lava. Several major volcanoes are erupting at any given moment, whereas on Earth there might not be as many major volcanic eruptions in a century. The largest volcanoes like Pele, named for the Hawaiian volcano goddess, eject volcanic material in huge umbrella-shaped plumes (Figure 137) that rise 150 miles or more and spread ejecta over areas as wide as 400 miles.

Io's volcanoes are not only more numerous but they are also more energetic than those on Earth. Material is ejected as though shot from a high-powered rifle at speeds of over 2000 miles per hour. This makes Io's volcanoes immensely more powerful than the most explosive volcanoes on Earth. Most of the large volcanoes, whose life spans are relatively short, only a few days or weeks, congregate in one hemisphere, while the smaller, more numerous volcanoes gather in the opposite hemisphere.

The large volcanoes spew out enormous quantities of lava that quickly obliterate any craters that form. A typical area on Io

Figure 136   View of Jupiter's moon Io from *Voyager 1*, showing numerous volcanic craters. Photo courtesy NASA

receives up to several inches of fresh lava each year. The larger volcanoes pour out hot sulfur lava, while the smaller ones eject sulfur dioxide, which quickly freezes, producing white snow-capped volcanic terrain. The tallest volcanoes are as high as Mount Everest, and since sulfur is too weak to support such huge structures, they are probably made of silicate rock, produced by volcanic eruptions similar to those on Earth.

Figure 137   A large volcanic plume rising from the surface of Io, the most volcanically active body in the Solar System. Photo courtesy USGS

SATURNIAN TECTONICS

Saturn, which is similar in size and composition to Jupiter but has only a third of its mass, possesses the strangest set of moons in the solar system (Figure 138). There are 17 in all, ranging in size from an asteroid to larger than Mercury. All but the outer two moons have orbits that are nearly circular, lie in the equatorial plane (the same plane as the rings), and keep the same side facing their mother planet, as does the Earth's moon.

Saturn's moons have densities less than twice the density of water, indicating that they are mostly composed of ice. For the majority of Saturn's moons the composition is 30 to 40 percent rock and 70 to 60 percent ice by weight. A surface reflectance, or albedo, between 60 and 90 percent for most of the moons suggests that they are coated with ice, making them highly reflective. With an albedo of nearly 100 percent, Enceladus, the second major moon outward from Saturn, is the most reflective body in the Solar System.

Iapetus, the second outermost moon, is half black and half white, which makes it disappear east of Saturn and reappear west of the planet. The dark side might have been formed by volcanic extrusions composed not of hot lava but of a slurry of ammonia, ice, and a dark material possibly organic in origin.

Rhea, the second largest of Saturn's moons has a densely cratered surface similar to the highlands on Mercury and the Earth's moon. Sharing much

Figure 138 Saturn and its moons as seen from *Voyager 1* in November 1980. Photo courtesy NASA

of the same terrain features as Rhea, Dione is also Saturn's second densest moon. Tethys has a branching canyon, 600 miles long, 60 miles wide, and several miles deep, spanning the distance between the North and South Polar regions. It also has a giant impact crater that is more than two-fifths the diameter of the moon itself.

Titan, which is larger than Mercury, is the only moon in the Solar System to possess a substantial atmosphere. Moreover, its atmosphere is even denser than the Earth's. It is composed of compounds of nitrogen, carbon, and hydrogen and is believed to resemble the Earth's atmosphere during its infancy. This makes Titan possibly the best place in the Solar System to look for the precursors of life. It is also the only known body in the Solar System besides the Earth whose surface is partially covered by a liquid, although its oceans are composed of liquid methane at temperatures of -175 degrees Celsius and its continents are made of ice.

URANIAN TECTONICS

Most of the moons of Jupiter, Saturn, and Uranus look as though they were resurfaced, cracked, and modified by flows of solid ice. Yet the moons seem to lack the internal energy sources needed to make these extensive changes. Oberon and Titania, the largest and outermost of the Uranian moons, are both a little less than half the size of the Earth's moon. Their surfaces, which are fairly uniform gray, are rich in water ice. Bright rays, assumed to be clean buried ice brought to the surface, shoot out from around several meteorite craters.

Oberon has a few features that look like faults, but it shows no evidence of geologic activity. Its surface is saturated with large craters, some more than 60 miles across. On the floor of several of the larger craters, volcanic activity has spewed out a mixture of ice and carbonaceous rock from the moon's interior.

The surface of Titania bears strong evidence of tectonic activity, including a complex set of rift valleys bounded by extensional faults, where the crust is being pulled apart (Figure 139). Titania's surface was also heavily cratered in its early history. However, many of the larger craters were erased when the moon was resurfaced by volcanism, which spewed water onto the crust. Some of the large craters have disappeared because they were flooded with water or because the soft icy crust collapsed. When the water in the interior began to freeze and expand, the entire surface stretched. Meanwhile, the crust ripped open, enormous blocks of ice dropped down along the faults, and water upwelled through the cracks to form smooth plains.

Images taken by *Voyager 2* in 1986 of Ariel and Miranda reveal evidence of solid-ice volcanism never before observed in the Solar System (Figure 140). A canyon like feature on Ariel called Brownie Chasma resembles a graben, which is a crack in the crust formed when the surface is pulled apart, causing large blocks to drop downward. The walls of the chasm are about 50 miles across, and the chasm floor bulges up into a round-topped ridge about a mile high.

Ariel's density suggests that its composition is mostly ice. However, the material

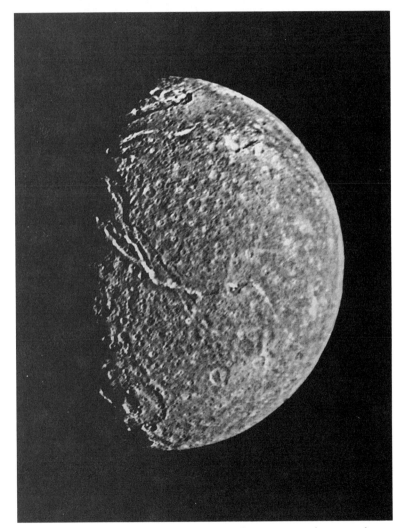

Figure 139 Titania, the largest Uranian moon, shows a large trenchlike feature near the day-night boundary which suggests tectonic activity. Photo courtesy NASA

rising up from the chasm floor cannot be liquid water because this would form a flat surface before it froze. Furthermore, the moon appears to lack any energy sources to maintain water in a liquid state. The material oozing up through the cracks would therefore have to be a plastic form of ice that piles up into a ridge instead of simply running out over the surface.

With a surface temperature of -200 degrees Celsius, the viscous flows do not travel very far and flatten out before they freeze. The presence of substances such as ammonia, methane, and carbon dioxide in ice buried beneath the surface could make it more buoyant than the surrounding rock-ice mixture. This would make the ice rise up through the cracks and erupt on the surface, producing a landscape seen nowhere else in the Solar System.

TRITON TECTONICS

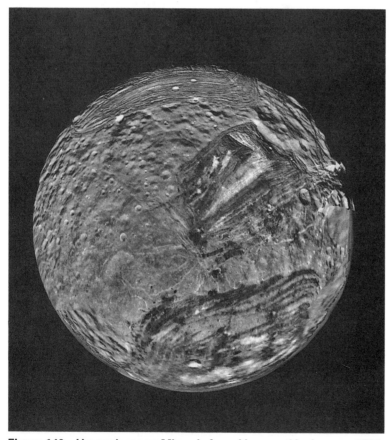

Figure 140   Uranus's moon Miranda from *Voyager 2* in January 1986, showing the strangest terrain features of any moon in our Solar System. Photo courtesy NASA

Triton, the largest moon of Neptune, appears to have been captured because its orbit is tilted 21 degrees to Neptune's equator and in a retrograde, or backward, direction, the only large moon in the Solar System known to do so. Triton also turns out to be the second most volcanically active body in the Solar System behind Jupiter's moon Io (followed by Venus and the Earth).

Triton's internal heat has long since vanished, so it was a difficult task to explain what powered its volcanic eruptions. Images sent back by *Voyager 2* in August 1989 showed dark plumes that seemed to indicate highly active volcanism on a body that is thought to have been tectonically dead for 4 billion years. Apparently,

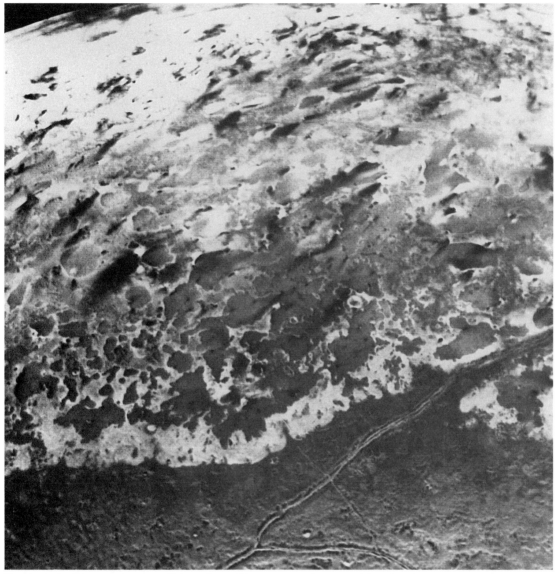

Figure 141   Neptune's moon Triton from *Voyager 2* in August 1989, showing evidence of extensive gas eruptions. Photo courtesy NASA

gigantic nitrogen-driven geysers spew out fountains of particles high into the thin, cold atmosphere and are strewn across the surface (Figure 141). As on Ariel, icy slush oozes out of huge fissures, and ice lava forms vast frozen lakes, providing a bewildering landscape as Voyager 2 takes a final glance before departing the Solar System on its journey through interstellar space.

175

GLOSSARY

abyssal	the deep ocean, generally over a mile in depth
accretion	the accumulation of celestial dust by gravitational attraction into a planetesimal, asteroid, moon, or planet
aerosol	a mass, made of solid or liquid particles dispersed in air
albedo	the amount of sunlight reflected from an object
asthenosphere	a layer of the upper mantle, roughly between 50 and 200 miles below the surface, which is more plastic than the rock above and below and might be in convective motion
astroblem	eroded remains on Earth's surface of an ancient impact structure produced by a large cosmic body
atmosphere	the gaseous envelope surrounding the Earth; the air
basalt	a volcanic rock that is dark in color and usually quite fluid in the molten state
basement	the surface beneath which sedimentary rocks are not found; the igneous, metamorphic, granitized or highly deformed rock underlying sedimentary rocks
batholith	the largest of intrusive igneous bodies, more than 40 square miles on its uppermost surface
biomass	the total mass of living organisms within a specified habitat

biosphere	the living portion of the Earth that interacts with all other geologic and biologic processes
block fault	type of normal fault in which the crust is separated into structural units (fault blocks) of different orientations and elevations
blueschists	metamorphosed rocks consisting of subducted ocean crust shoved up on continents by plate tectonics
caldera	a large pitlike depression found at the summits of some volcanoes that is formed by great explosive activity and collapse
catastrophism	a theory that ascribes to the belief that recurrent, violent, worldwide events are the reason for the sudden disappearance of some species and the abrupt rise of new ones
chert	a dense, extremely hard siliceous sedimentary rock, consisting mainly of interlocking quartz crystals
chondrite	the most common type of meteorite, composed mostly of rocky material with small spherical grains
circumpolar	traveling around the world from pole to pole
comet	a celestial body believed to come from a cloud of comets that surrounds the sun and develops a long tail of gas and dust particles when traveling near the inner Solar System
conductivity	the property of transmitting a quality
continent	a landmass composed of light, granitic rock riding on denser rocks of the upper mantle
continental shelf	the offshore area of a continent in shallow sea
correlation	the determination of geological age by comparing fossils from different times and regions
craton	the stable interior of a continent, usually composed of the oldest rocks on the continent, commonly Precambrian
Cretaceous-Tertiary boundary	see K-T boundary.

crosscutting	a body of rocks cutting across older rock units
daughter product	a product formed from another through radioactive decay
diapir	a vertical columnar plug of less dense rock or magma that is forced through a more dense rock
diatom	any of numerous microscopic unicellular marine or freshwater algae having siliceous cell walls
diatomite	an ultra-fine-grained siliceous earth composed mainly of diatom cell walls
dike	a body of intrusive igneous rock that cuts across the layering or structural fabric of the host rock
dolomite	see dolomitization.
dolomitization	the process by which limestone becomes dolomite by the substitution of magnesium for the original calcite; common in organisms whose original hard parts were composed of calcite or aragonite, such as corals, brachiopods, and echinoderms
down-drop	the lowering of a fault block
downfault	the down-dropping of the crust along a fault plane
dynamo effect	the creation of the Earth's magnetic field by rotational, thermal, chemical, and electrical differences between the solid inner core and the liquid outer core
eon	on the geologic time scale, the longest unit of time, comprised of several eras
epoch	an interval of geologic time longer than an age and shorter than a period
era	on the geologic time scale, the unit of time below eon; comprised of several periods
erratic	a glacially deposited boulder far from its source
escape velocity	the minimum velocity that an object must have in order to escape from the surface of a planet (or moon) against the gravitational attraction

extrusive	magma pouring out onto the Earth's surface as lava and ash
fault	a breaking of crustal rocks caused by Earth's movements
fault-block mountain	a mountain formed by a block faulting, i.e., isolated by faulting and categorized as structural or tectonic
gabbro	a group of crystalline intrusive rocks composed chiefly of plagioclase and pyroxene
Genesis Rock	ancient moon rocks forming the original lunar crust
geochemical	pertaining to the distribution and circulation of chemical elements in the Earth's soil, water, and atmosphere
geodetic	pertaining to the study of the external shape of the Earth as a whole
geologic column	a diagram showing in columns the total succession of stratigraphic units of a region
geosphere	the inorganic world, including the lithosphere (solid portions of the Earth), the hydrosphere (bodies of water), and the atmosphere (air)
geosyncline	a basinlike or elongated subsidence of the Earth's crust. Its length may extend for several thousand miles and may contain sediments thousands of feet thick, representing millions of years of deposits. A geosyncline generally forms along continental edges and is destroyed during periods of crustal deformation.
geothermal	the generation of hot water or steam by hot rocks in the Earth's interior
gneiss	a banded, coarse-grained metamorphic rock with alternating layers of unlike minerals. It consists of essentially the same components as granite.
graben	a crustal unit or block that has been depressed relative to the blocks on either side
granitic intrusion	the injection of magma into the crust that forms granitic rocks
greenhouse effect	the trapping of heat in the atmosphere by carbon dioxide

greenhouse gases   gases, principally water vapor and carbon dioxide, that trap surface heat, thereby warming the Earth

guyot   a seamount of more or less circular form and having a flat top, thought to be volcanic cones whose tops have been flattened by surface-wave action

high-angle fault   a fault with a dip exceeding 45 degrees

hot spot   a heat source deep within the Earth's mantle, persistent for millions of years

hydrologic cycle   the constant interchange of water between the oceans, atmosphere, and land areas of the Earth by way of transpiration and evaporation

hydrosphere   the water layer at the surface of the Earth

hypocenter   the point of origin of earthquakes, also called the focus

igneous   rock that comprises the Earth's crust, formed of hardened magma

insolation   all solar radiation impinging on the Earth

intrusive   any igneous body which has solidified in place below the surface of the Earth. Compare with extrusive.

iridium   a rare isotope of platinum, relatively abundant on meteorites

island arc   a curved chain of volcanic islands, many of which are located along the circum-Pacific margins. Island arcs are commonly associated with deep-sea trenches.

isostasy   the equilibrium achieved by blocks of crustal rock floating on denser rocks below

isotope   a particular atom of an element that has the same number of electrons and protons as the other atoms of the element, but a different number of neutrons; i.e., the atomic numbers are the same, but the atomic weights differ

kimberlite   a diamond-bearing volcanic material originating deep within the mantle

kimberlite pipes diamond bearing pipelike structures in South Africa. The pipes are known as diatremes, and are formed deep within the Earth's crust by the explosion of magmatic gases or heated groundwater.

K-T boundary the boundary formation marking the end of the dinosaur era

leach (out) the dissolution of soluble substances in rocks by the percolation of meteoric water

limestone a sedimentary rock composed of calcium carbonate that is secreted from seawater by invertebrates whose skeletons composed the bulk of deposits

lithosphere a rigid outer layer of the mantle, typically about 60 miles thick. It is overridden by the continental and oceanic crusts and is divided into segments called plates.

lithospheric plate one of several separate segments of the lithosphere. Plate tectonics involves the interactions of lithospheric plates.

lystrosaurus an ancient mammal-like reptile

magma a molten rock material generated within the Earth, it is the constituent of igneous rocks, including volcanic eruptions

mantle the part of the Earth below the crust and above the core, composed of dense iron-magnesium-rich rocks

maria dark plains on the lunar surface caused by massive basalt floods

marsupials primitive, warm-blooded animals that wean their young in a belly pouch

mass a measure of the amount of matter in a body

massive sulfides sulfide metals deposited from hydrothermal solutions

metamorphic rock a rock crystallized from previous igneous, metamorphic, or sedimentary rocks created under conditions of intense temperatures and pressures without melting

meteorite a metallic or stony body from space that enters the Earth's atmosphere and impacts on the Earth's surface

microspherule	tiny glassy beads generated by a large meteorite impact
Mohorovicic discontinuity/ Moho	the boundary between the crust and mantle, discovered by Andrija Mohorovicic
natural selection	the process by which evolution selects species for survival or extinction depending on the environment
Neopangaea	a hypothetical future supercontinent formed when the present continents collide with each other
neptunism	the belief that rocks precipitated from seawater
net slip	the distance between two previously adjacent points on either side of a fault, it defines both the direction and size of displacement
nontransform offset	small offsets with overlapping ridge tips at the Mid-Atlantic Ridge
ophiolite	masses of igneous rocks, such as basalt, whose structure and composition identify them as segments of ocean crust pushed into the continents by plate collisions
original horizontality	the theory that rocks were originally deposited in the ocean horizontally and later folding and faulting tilted the beds
orogens	eroded roots of ancient mountain ranges
orogeny	an episode of mountain building
ozone	a molecule consisting of three atoms of oxygen that exists in the upper atmosphere and filters out ultraviolet light from the sun
ozone layer	a diffused layer of ozone between 20 and 30 miles altitude that shields the Earth from damaging ultraviolet radiation from the sun
Pangaea	a supercontinent that existed about 300 to 200 mya, and included most of the continental crust of the Earth
Panthalassa	the ocean that surrounded Pangaea before its fragmentation

period   a division of geologic time longer than an epoch and included in an era

photosynthesis   the process by which plants create carbohydrates from carbon dioxide, water, and sunlight

pillow lava   spherical structures usually composed of basaltic lava, generally about 3 feet in diameter. These formations are the result of the rapid cooling of hot, fluid magma that comes in contact with water.

plate tectonics   the theory that accounts for the major features of the Earth's surface in terms of the interaction of lithospheric plates. See tectonics.

plutonism   the belief that rocks formed from molten material from the Earth's interior

polarity   a condition in which a substance exhibits opposite properties such as electrical charges or magnetism

precipitation   products of condensation that fall from clouds as rain, snow, hail, or drizzle; also the deposition of rocks from seawater

pyroclastic   the fragmental ejecta released explosively from a volcanic vent

radioactive decay   the process by which radioactive isotopes decay into stable elements

radioactivity   an atomic reaction releasing detectable radioactive particles

radiogenic   pertaining to something produced by radioactive decay, such as heat

radiometric dating   a method of dating fossil remains by the testing of stable versus unstable radioactive material

regolith   unconsolidated rock material resting on bedrock, found at and near the Earth's surface

residence time   the time required for a factor to remain in a certain environment; for example, carbon dioxide in the ocean

rift   the center of an extensional spreading center, where continental or oceanic plate separation occurs

rock bed	a rock unit distinguishable from beds beneath or above it
rock correlation	the tracing of equivalent rock exposures over distance
seafloor spreading	the theory that the ocean floor is created by the separation of lithospheric plates along the midocean ridges, with new oceanic crust formed from mantle material that rises from the mantle to fill the rift
seamount	an isolated submarine elevation of at least 2300 feet (700 meters)
sediment	the debris, organic or inorganic, transported and deposited by wind, water, or ice. It may form loose sediment, like sand or mud, or become consolidated to form sedimentary rock.
sedimentation	the process by which sediment is formed, transported, and deposited
seismic	pertaining to earthquakes
shield	areas of the exposed Precambrian nucleus of a continent
sial	a lightweight layer of rock that lies below the continents
sill	a tabular igneous intrusion with boundaries parallel to the planar structure of the surrounding rock
sima	a dense rock that composes the ocean floor and on which the sial floats
spherules	small, spherical, glassy grains found on certain types of meteorites, lunar soils, and large meteorite impact sites on Earth
stock	an intrusive body of deep-seated igneous rock, usually lacking conformity and resembling a batholith, except for its smaller size
strata	layered rock formations; also called beds
stratification	a pattern of layering in sedimentary rock, lava flows or water, or materials of different composition or density

stromatolite	sedimentary formation in the shape of cushions or columns produced by lime-secreting blue-green algae (cyanobacteria)
subduction zone	an area where the oceanic plate dives below a continental plate into the asthenosphere. Ocean trenches are the surface expression of a subduction zone.
succession	a sequence of rock formations
superposition, law of	in any sequence of sedimentary strata that is not strongly folded or tilted, the youngest strata is at the top and the oldest at the bottom
tectonics	in geology, the history of the larger features of the earth (rock formations and plates) and the forces and movements that produce them. See plate tectonics.
terrain	a region of the Earth's surface that is treated as a physical feature or as a type of environment
terrestrial planet	the rocky planets Mercury, Venus, Earth, and Mars
thermal conductivity	the amount of heat conducted per unit of time through any cross section of a substance, dependent on the temperature gradient at that section and the area of the section
till	nonstratified material deposited directly by glacial ice as it recedes
tillite	a sedimentary rock formed by the compaction and cementation of till
transform fault	a fracture in the Earth's crust along which lateral movement occurs. They are common features of the midocean ridges, created in the line of seafloor spreading.
transgression	the flow of the sea over land areas, or a change that converts initially shallow-water conditions to deep-water conditions
trench	a depression on the ocean floor caused by subduction
troposphere	the lowest 6 to 12 miles of the Earth's atmosphere, characterized by decreasing temperature with increasing height

ultramafic — an igneous rock rich in iron and magnesium and poor in silica

ultraviolet radiation — the invisible light with a wavelength shorter than visible light and longer than x-rays

unconformity — a lack of continuity in the stratigraphic record, caused by a weathering or erosion of surface before new layers are deposited

uniformitarianism — the belief that the slow processes that shape the Earth's surface have acted essentially unchanged throughout geologic time

upfault — a block of crust pushed upwards along fault lines

volatiles — in the magma, those materials that readily form a gas and are the last to enter into and crystallize as minerals during solidification

BIBLIOGRAPHY

CONTINENTAL DRIFT

Ballard, Robert D. *Exploring Our Living Planet*. Washington, D.C.: National Geographic Society, 1983.

Briggs, Peter. *Mysteries of Our World*. New York: McKay, 1969.

Golden, Frederic. *The Moving Continents*. New York: Scribners, 1972.

Harrington, John W. *Dance of the Continents*. New York: J. P. Tarcher, 1983.

Hurley, P. M. *Advances in Earth Science*. Cambridge: MIT Press, 1964.

Marvin, Ursula B. *Continental Drift; The Evolution of a Concept*. Washington, D.C.: Smithsonian Press, 1973.

Miller, Russell. *Continents in Collision*. Alexandria, Va.: Time-Life Books, 1983.

Parker, Ronald B. *Inscrutable Earth; Explorations into the Science of Earth*. New York: Scribners, 1984.

Simon, Cheryl. "The Great Earth Debate." *Science News* 121 (March 13, 1982): 178–179.

Weiner, Jonathan. *Planet Earth*. New York: Bantam, 1986.

HISTORIC TECTONICS: AN OVERVIEW

Barron, Eric J., and William H. Peterson. "Model Simulation of the Cretaceous Ocean." *Science* 244 (May 12, 1989): 684–686.

Boucot, A. J., and Jane Gray. "A Paleozoic Pangaea." *Science* 222 (November 11, 1983): 571–580.

Gambles, Peter. "Death of an Ancient Ocean." *Nature* 312 (November 1984): 400–401.

Kerr, Richard A. "Another Movement in the Dance of the Plates." *Science* 244 (May 5, 1989): 529–530.

———. "Plate Tectonics Is the Key to the Distant Past." *Science* 234 (November 7, 1986): 670–672.

———. "Plate Tectonics Goes Back 2 Billion Years." *Science* 230 (December 20, 1985): 1364–1367.

Krogstad, E. J., et al. "Plate Tectonics 2.5 Billion Years Ago: Evidence at Kolar, South India." *Science* 243 (March 10, 1989): 1337–1339.

Kunzig, Robert. "Horizontal History." *Discover* 10 (September 1989): 16–18.

Monastersky, Richard. "New Record for World's Oldest Rocks." *Science News* 136 (October 7, 1989): 228.

Simon, Cheryl. "Deep Crust Hints Meteoritic Impact." *Science News* 121 (January 30, 1982): 69.

CONVECTION CURRENTS

Anderson, Don L., and Adam M. Dziewonski. "Seismic Tomography." *Scientific American* 251 (October 1984): 60–68.

Bercovici, Dave, Gerald Schubert, and Gary A. Glatzmaier. "Three-Dimensional Spherical Models of Convection in the Earth's Mantle." *Science* 244 (May 26, 1989): 950–954.

Bonatti, Enrico. "The Rifting of Continents." *Scientific American* 256 (March 1987): 97–103.

Courtillot, Vincent, and Gregory E. Vink. "How Continents Break Up." *Scientific American* 249 (July 1983): 43–49.

Heppenheimer, T. A. "Journey to the Center of the Earth." *Discover* 8 (November 1987): 86–93.

Howell, David G. "Terranes." *Scientific American* 253 (November 1985): 116–125.

Monastersky, Richard. "Spinning the Supercontinent Cycle." *Science News* 135 (June 3, 1989): 344–346.

Nance, R. Damian, Thomas R. Worsley, and Judith B. Moody. "The Supercontinent Cycle." *Scientific American* 259 (July 1988): 72–79.

O'Nions, R. K., P. J. Hamilton, and Norman M. Evensen. "The Chemical Evolution of the Earth's Mantle." *Scientific American* 242 (May 1980): 120–133.

White, Robert S., and Dan P. McKenzie. "Volcanism at Rifts." *Scientific American* 261 (July 1989): 62–71.

BIBLIOGRAPHY

CRUSTAL PLATES

Burke, Kevin C., and J. Tuzo Wilson. "Hot Spots on Earth's Surface." *Scientific American* 235 (August 1976): 46–57.

Carter, William E., and Douglas S. Robertson. "Studying the Earth by Very-Long-Baseline Interferometry." *Scientific American* 255 (November 1986): 46–54.

Johnston, Arch C., and Lisa R. Kanter. "Earthquakes in Stable Continental Crust." *Scientific American* 262 (March 1990): 68–75.

Jordan, Thomas H., and J. Bernard Minster. "Measuring Crustal Deformation in the American West." *Scientific American* 259 (August 1988): 48–58.

Kerr, Richard A. "Slices of Continental Crust Coming into View." *Science* 242 (October 14, 1988): 195–197.

Monastersky, Richard. "Set Adrift by Wandering Hotspots." *Science News* 132 (October 17, 1987): 250–252.

Stein, Ross S., and Robert S. Yeats. "Hidden Earthquakes." *Scientific American* 260 (June 1989): 48–57.

Vink, Gregory E., W. Jason Morgan, and Peter R. Vogt. "The Earth's Hot Spots." *Scientific American* 252 (April 1985): 50–57.

Weisburd, Stefi. "Seeing Continents Drift." *Science News* 128 (December 21, 1985): 388.

SEAFLOOR SPREADING

Bonatti, Enrico, and Kathleen Crane. "Ocean Fracture Zones." *Scientific American* 250 (May 1984): 40–51.

Edmond, John M., and Karen Von Damm. "Hot Springs on the Ocean Floor." *Scientific American* 248 (April 1983): 78–93.

Hekinian, Roger. "Undersea Volcanoes." *Scientific American* 251 (July 1984): 46–55.

Hoffman, Kenneth A. "Ancient Magnetic Reversals: Clues to the Geodynamo." *Scientific American* 258 (May 1988): 76–83.

Kerr, Richard A. "Sea-Floor Spreading Is Not So Variable." *Science* 223 (February 3, 1984): 472–473.

Monastersky, Richard. "Mid-Atlantic Ridge Survey Hits Bull's-eye." *Science News* 135 (May 13, 1989): 295.

Mutter, John C. "Seismic Images of Plate Boundaries." *Scientific American* 254 (February 1986): 66–75.

Sclater, John G., and Christopher Tapscott. "The History of the Atlantic." *Scientific American* 240 (June 1979): 156–174.

Yulsman, Tom. "Plate Tectonics Revised." *Science Digest* 93 (November 1985): 35.

SUBDUCTION ZONES

Bartusiak, Marcia. "Mapping the Sea Floor from Space." *Popular Science* 224 (February 1984): 81–85.

Bernardo, Stephanie. "The Seafloor: A Clear View from Space." *Science Digest* 92 (June 1984): 44–48.

Francheteau, Jean. "The Ocean Crust." *Scientific American* 249 (September 1983): 114–129.

Frohlich, Cliff. "Deep Earthquakes." *Scientific American* 260 (January 1989): 48–55.

Heaton, Thomas H., and Stephen H. Hartzell. "Earthquake Hazards on the Cascadia Subduction Zone." *Science* 236 (April 10, 1987): 162–168.

Monastersky, Richard. "Birth of a Subduction Zone." *Science News* 136 (December 16, 1989): 396.

———. "Catching Subduction in the Act." *Science News* 133 (January 2, 1988): 8.

Peacock, Simon M. "Fluid Processes in Subduction Zones." *Science* 248 (April 20, 1990): 329–336.

MOUNTAIN BUILDING

Armstead, Christopher H. *Geothermal Energy*. New York: Wiley, 1978.

Bird, Peter, "Formation of the Rocky Mountains, Western United States: A Continuum Computer Model." *Science* 239 (March 25, 1988): 1501–1507.

Cook, Frederick A., Larry D. Brown, and Jack E. Oliver. "The Southern Appalachians and the Growth of the Continents." *Scientific American* 243 (October 1980): 156–168.

Gass, Ian G. "Ophiolites." *Scientific American* 247 (August 1982): 122–131.

Kerr, Richard A. "Making Mountains with Lithospheric Drips." *Science* 239 (February 26, 1988): 978–979.

Molnar, Peter. "The Structure of Mountain Ranges." *Scientific American* 255 (July 1986): 70–79.

Talbot, Christopher, J., and Martin P. A. Jackson. "Salt Tectonics." *Scientific American* 257 (August 1987): 70–79.

BIBLIOGRAPHY

THE ROCK CYCLE

Berner, Robert A., and Antonio C. Lasaga. "Modeling the Geochemical Carbon Cycle." *Scientific American* 260 (March 1989): 74–81.

Broecker, Wallace S. "Carbon Dioxide Circulation through Ocean and Atmosphere." *Nature* 308 (April 12, 1984): 602.

Cathles, Lawrence M. III. "Scales and Effects of Fluid Flow in the Upper Crust." *Science* 248 (April 20, 1990): 323–328.

Kerr, Richard A. "The Carbon Cycle and Climate Warming." *Science* 222 (December 9, 1983): 1107–1108.

Monastersky, Richard. "The Whole-Earth Syndrome." *Science News* 133 (June 11, 1988): 378–380.

Schneider, Stephen H. "Climate Modeling." *Scientific American* 256 (May 1987): 72–80.

Siever, Raymond. "The Steady State of the Earth's Crust, Atmosphere, and Oceans." *Scientific American* 230 (June 1974): 72–79.

TECTONICS AND LIFE

Cloud, Preston. "The Biosphere." *Scientific American* 249 (September 1983): 176–189.

Hallam, Anthony. "End-Cretaceous Mass Extinction Event: Argument for Terrestrial Causation." *Science* 238 (November 27, 1987): 1237–1241.

Kerr, Richard A. "Was There a Prelude to the Dinosaurs' Demise?" *Science* 239 (February 12, 1988): 729–730.

Macdonald, Ken C., and Bruce P. Luyendyk. "The Crest of the East Pacific Rise." *Scientific American* 224 (May 1981): 100–116.

Morris, S. Conway. "Burgess Shell Faunas and the Cambrian Explosion." *Science* 246 (October 20, 1989): 339–345.

Officer, Charles B., and Charles L. Drake. "The Cretaceous-Tertiary Transition." *Science* 219 (March 25, 1983): 1383–1390.

Rampino, Michael R., and Richard B. Strothers. "Flood Basalt Volcanism during the Past 250 Million Years." *Science* 241 (August 5, 1988): 663–667.

Stanly, Steven M. "Mass Extinctions in the Ocean." *Scientific American* 250 (June 1984): 64–72.

Valentine, James W., and Eldridge M. Moors. "Plate Tectonics and the History of Life in the Oceans." *Scientific American* 230 (April 1974): 80–89.

Weisburd, Stefi. "Volcanoes and Extinctions: Round Two." *Science News* 131 (April 18, 1987): 248–250.

Tectonics in Space

Carr, Michael H. "The Volcanoes of Mars." *Scientific American* 234 (January 1976): 33–43.

Croswell, Ken. "Io, Jupiter's Fiery Satellite." *Space World* 295 (July 1988): 12–15.

Eberhart, Jonathan. "Solid Ice Volcanism on Uranian Moons." *Science News* 134 (September 17, 1988): 183.

Johnson, Torrence V., Robert Hamilton Brown, and Laurence A. Soderblom. "The Moons of Uranus." *Scientific American* 256 (April 1987): 48–60.

Kerr, Richard A. "Triton Steals *Voyager's* Last Show." *Science* 245 (September 1, 1989): 928–930.

———. "Venus is Looking More Like Earth than Mars." *Science* 232 (May 9, 1986): 709–710.

Lucchitta, B. K. "Recent Mafic Volcanism on Mars." *Science* 235 (January 30, 1987): 565–567.

Pettengill, Gordon H., Donald B. Campbell, and Harold Masursky. "The Surface of Venus." *Scientific American* 243 (August 1980): 54–65.

Prinn, Ronald G. "The Volcanoes and Clouds of Venus." *Scientific American* 252 (March 1985): 46–53.

Soderblom, Laurence A. "The Galilean Moons of Jupiter." *Scientific American* 242 (January 1980): 88–100.

Soderblom, Laurence A., and Torrence V. Johnson. "The Moons of Saturn." *Scientific American* 246 (January 1982): 101–116.

Taylor, S. R. "Source of the Oldest Lunar Basalt." *Nature* 310 (July 12, 1984): 98–99.

INDEX